BOOTH RENTING
A Guide for the Independent Stylist

Australia • Brazil • Japan • Korea • Mexico • Singapore • Spain • United Kingdom • United States

**Booth Renting 101:
A Guide for the Independent Stylist**

Executive Director, Milady: Sandra Bruce

Product Director: Corina Santoro

Product Manager: Philip I. Mandl

Product Team Manager: Julie Shepperly

Senior Content Developer: Jessica Mahoney

Associate Content Developer: Sarah Prediletto

Director, Marketing & Training: Gerard McAvey

Community Manager: Matthew McGuire

Senior Production Director: Wendy Troeger

Production Manager: Sherondra Thedford

Senior Content Project Manager: Stacey Lamodi

Senior Art Director: Benj Gleeksman

Cover image(s): © iStockPhoto.com/stask

© 2015 Milady, a part of Cengage Learning

WCN: 01-100-101

ALL RIGHTS RESERVED. No part of this work covered by the copyright herein may be reproduced, transmitted, stored, or used in any form or by any means graphic, electronic, or mechanical, including but not limited to photocopying, recording, scanning, digitizing, taping, Web distribution, information networks, or information storage and retrieval systems, except as permitted under Section 107 or 108 of the 1976 United States Copyright Act, without the prior written permission of the publisher.

For product information and technology assistance, contact us at
Cengage Learning Customer & Sales Support, 1-800-354-9706

For permission to use material from this text or product,
submit all requests online at **www.cengage.com/permissions**.
Further permissions questions can be e-mailed to
permissionrequest@cengage.com

Library of Congress Control Number: 2012952890

ISBN-13: 978-1-2850-6327-0

Cengage Learning
5 Maxwell Drive
Clifton Park, NY 12065-2919
USA

Cengage Learning is a leading provider of customized learning solutions with office locations around the globe, including Singapore, the United Kingdom, Australia, Mexico, Brazil, and Japan. Locate your local office at:
international.cengage.com/region

Cengage Learning products are represented in Canada by Nelson Education, Ltd.

For your lifelong learning solutions, visit **milady.cengage.com** Purchase any of our products at your local college store or at our preferred online store **www.cengagebrain.com**

Visit our corporate website at **cengage.com**.

Notice to the Reader
Publisher does not warrant or guarantee any of the products described herein or perform any independent analysis in connection with any of the product information contained herein. Publisher does not assume, and expressly disclaims, any obligation to obtain and include information other than that provided to it by the manufacturer. The reader is expressly warned to consider and adopt all safety precautions that might be indicated by the activities described herein and to avoid all potential hazards. By following the instructions contained herein, the reader willingly assumes all risks in connection with such instructions. The publisher makes no representations or warranties of any kind, including but not limited to, the warranties of fitness for particular purpose or merchantability, nor are any such representations implied with respect to the material set forth herein, and the publisher takes no responsibility with respect to such material. The publisher shall not be liable for any special, consequential, or exemplary damages resulting, in whole or part, from the readers' use of, or reliance upon, this material.

Printed in the United States of America
1 2 3 4 5 6 7 18 17 16 15 14

brief contents

Table of Contents   iv
About the Author   vi
Preface   viii
Reviewers   x

CHAPTER 1
The Basics of Running a Booth-Rental Operation 1

CHAPTER 2
Getting Your Business Off the Ground and Running 29

CHAPTER 3
Eliminate Detours and Distractions with a Solid Business Model 61

CHAPTER 4
Your Money, Your Future 95

CHAPTER 5
Marketing Your Booth-Rental Business 119

CHAPTER 6
Strategies for Retention and Building Clientele. 141

CHAPTER 7
The Day-to-Day Details. 155

Appendix A: Self-Assessment 178
Appendix B: Answers to End-of-Chapter Quizzes 180
Glossary 186
Index 190

table of contents

About the Author vi

Preface viii

Reviewers x

1 The Basics of Running a Booth-Rental Operation / 1

- 1.1 How to Decide Whether Booth Rental Is Right for You / 4
- 1.2 Your Responsibilities as a Booth Renter / 11
- 1.3 Summary / 21
- 1.4 Top Takeaways: Basics of Running a Booth-Rental Operation / 21

2 Getting Your Business Off the Ground and Running / 29

- 2.1 Choosing a Business Structure / 31
- 2.2 Licenses / 38
- 2.3 Finding the Ideal Location to Rent a Booth / 39
- 2.4 Work and Sublease Agreements / 46
- 2.5 Understanding Your Legal Rights as a Booth Renter / 48
- 2.6 Summary / 48
- 2.7 Top Takeaways: Getting Your Business Off the Ground and Running / 49

3 Eliminate Detours and Distractions with a Solid Business Model / 61

- 3.1 Create a Vision and Mission Statement / 67
- 3.2 Set Goals for Your Business / 68
- 3.3 Identify Your Core Business Values / 70
- 3.4 Create a Solid Business Plan / 74
- 3.5 Growing the Business / 80
- 3.6 S.W.O.T. Analysis / 88
- 3.7 Summary / 91
- 3.8 Top Takeaways: Creating a Solid Business Model / 92

4 Your Money, Your Future / 95

- 4.1 Costs to Succeed / 97
- 4.2 Evaluate Your Personal Expenses / 99
- 4.3 Establish Income Goals / 99
- 4.4 Establish Your Financial Dream Team / 101
- 4.5 Create a Financial-Health Checklist / 104
- 4.6 Analyze Your Financial Health / 106
- 4.7 Health Benefits / 108

	4.8	Taxes / 109
	4.9	Saving for Retirement / 112
	4.10	Build Business Credit / 113
	4.11	Summary / 115
	4.12	Top Takeaways: Your Money, Your Future / 115

Marketing Your Booth-Rental Business / 119

- 5.1 Understanding Your Brand / 121
- 5.2 Creating a Marketing Plan and Marketing Strategy / 122
- 5.3 Marketing and Promotion Strategies / 128
- 5.4 The Marketing and Promotions Calendar Template / 135
- 5.5 Summary / 137
- 5.6 Top Takeaways: Marketing Your Booth-Rental Business / 137

Strategies for Retention and Building Clientele / 141

- 6.1 Client Partnerships / 143
- 6.2 Establish Client Incentives / 149
- 6.3 Client Retention Strategies / 149
- 6.4 Coaching Corner: Business Ethics / 150
- 6.5 Summary / 151
- 6.6 Top Takeaways: Strategies for Retention and Building Clientele / 151

The Day-to-Day Details / 155

- 7.1 Scheduling Appointments / 157
- 7.2 Creating Services / 159
- 7.3 Retailing / 161
- 7.4 Pricing Your Retail / 162
- 7.5 Coaching Corner: Formula for Success / 164
- 7.6 Keeping Track of Your Product Inventory / 165
- 7.7 Accepting Payments / 167
- 7.8 Tracking Your Income / 168
- 7.9 Summary / 172
- 7.10 Top Takeaways: The Day-to-Day Details / 172

Appendix A: Self-Assessment / 178
Appendix B: Answers to End-of-Chapter Quizzes / 180
Glossary / 186
Index / 190

about the author

DESHAWN F. BULLARD

Originally from Chicago, Illinois, DeShawn Bullard has meshed her passion for hair, technology, and entrepreneurship to carve out a successful business niche in the cosmetology field. While earning a bachelor's degree in computer science from Tuskegee University, styling hair was a convenient way to make extra money. After graduation, DeShawn landed successfully in Corporate America as a computer programmer, where she worked for seven years. However, her true passion for hair kept calling, and she chose to follow her dream. While working as a computer programmer by day, she studied cosmetology by night and received her license from Atlanta Area Technical School. Her goal was not only to be the best stylist but to run the most professional business in the cosmetology field.

DeShawn has become known and recognized for her creative styling techniques, but is also known for her skills in educating, motivating, encouraging, and empowering beauty-industry professionals on how to take their business to the next level. Along with being a product manufacturer, she is an experienced platform artist who has conducted numerous workshops and seminars for such shows as the Bronner Brothers Trade Shows, International Hair Shows, the Chicago Midwest Beauty Show (in which she was selected as keynote speaker for their first multicultural conference), and many other trade shows throughout the world. Her dynamic style of teaching afforded her a position as Educational Director for Professional Products, Inc. As Educational Director, she traveled nationally and internationally to such places as Canada, Brazil, the Bahamas, and the Virgin Islands, educating stylists about the latest in hair care, business, and the use of technology in the salon industry. She has also excelled in the competition arena, winning such competitions as the Bronner Brothers Day by Night Competition and the Mr. Raw Male Texturizing Category.

Today, CEO DeShawn Bullard provides recession-proof products and services to consumers both nationally and internationally. As president of her company, she continues to broaden the distribution of NouriTress Perfect Hair Products, the company she founded in 1998. The popularity of her products allowed her to team up with Grammy Award–winning songstress Tionne "T-Boz" Watkins, who had become the face and national spokesperson for NouriTress Perfect Hair Products in 2008. Her products, which received an *Essence* magazine beauty award, are distributed nationally in major retail stores such as JCPenney and Target, with international distribution in places such as Canada, the United Kingdom, and Africa.

Nominated by the Atlanta Business League for "Outstanding Business Woman of the Year" in 2009, DeShawn is steadily raising the bar by operating several business entities, including the NouriTress Salon and Hair Clinic. DeShawn has made several appearances as a guest hair stylist on the Turner Broadcasting System's (TBS) nationally syndicated TV show *Movie and a Makeover*, and her work has graced the pages of top beauty-industry magazines like *Essence, Sophisticate's Hairstyle Guide, Beauty Store Business, Hype Hair*, and *American Salon*, just to name a few. For four years, she provided weekly hair advice to millions of listeners as she dished out hot hair tips on Sheridan Broadcasting Radio. For six consecutive years, she was named one of the most influential women in the beauty industry in *SalonSENSE magazine*. Top beauty companies like L'Oreal Mizani also called on her to lead educational tours, and Sallys Beauty selected her to create the look of their product line Silk Elements.

From London to Brazil, to Canada and the United States, beauty-industry professionals around the world pay attention when they hear her name. Her workshops sell out with standing-room-only at any beauty event, such as the Bronner Brothers Hair Show. DeShawn has shown the beauty industry how you can go from standing behind the chair servicing clients to working on top celebrity clientele, such as T-Boz, Lynn Whitfield, Ann Nesby, Tramaine Hawkins, and Betty Wright, all while building a multimillion-dollar hair product line.

preface

Welcome to the world of the independent beauty professional. The world of independence is wonderful, but there are some things you must know about being independent to make your business a success. Do you remember how it felt to leave your parents' house and become an independent adult, living on your own? It felt good and scary at the same time, but the time you spent at your parents' house living under their rules helped you prepare for your independence. Becoming an independent beauty professional is the same. The time you spent working in a salon for someone else was your training ground for becoming independent. Working in a salon as an employee taught you about time management, scheduling appointments, client retention, and mastering techniques. These are all tools you will have to use as an independent beauty professional, but there is so much more you have to learn.

This book was designed to assist beauty professionals who work independently as a booth renter, mobile contract professional, or in a home-based business to properly run their independent beauty business.

Cosmetology school is designed to teach you the basics about hair, makeup, skin, and nails, but it does not always teach you how to properly run an independent business. Upon completing cosmetology school, some professionals think about booth renting because they can set their own hours, set their own prices, and keep all the profits. Others have been working as an employee and are ready to move on to independence, while many have been functioning as an independent beauty professional for quite awhile and are running into some challenges in making this type of structure successful.

When you decide to become an independent beauty professional or booth renter, what you must understand is that you are officially running your own small business. When you open a small business, there are certain things required by law that you must have in place in order to run that business. The reason this book is written is many beauty-industry professionals have taken on the role of a booth renter and have no clue what their rights and responsibilities are, and struggle or sometimes fail while operating as an independent beauty professional.

After I graduated from college and worked in Corporate America for seven years as a computer programmer, I decided to give up the suit and heels for a career in the beauty industry. After graduation from beauty school, I immediately began working as a booth renter. Following a year of booth renting, I owned and operated a booth-rental salon, but had no real direction. During those years, I had no manual to tell me what to do, so I made many mistakes. I fell, but I got back up again, and kept pushing until I succeeded, because this is what I had learned from my corporate experiences. Once I mastered the art of booth renting, I decided to help others who faced the same challenges I had. I trained and mentored beauty-industry professionals who became booth renters too early in their careers, and who got lost while trying to run this type of structure because they had nowhere to get information about how to properly run and operate a booth-rental business.

This book provides a roadmap for licensed independent beauty professionals on how to properly start and successfully run an independent beauty business. Its chapters are filled with information, forms, and exercises that will give those seeking information about booth renting the tools they need to be successful. At the end of each chapter, you will find "Top Takeaways"—checklists, charts, and informational resources—that will help you take your business to the next level. While reading this book, I encourage you to think big and look at the big picture. Take one step at a time until you accomplish your goal. If you think it and believe it, you can achieve it!

NOTE: In this book, you will notice terms such as *independent beauty professional*, *independent stylist*, and *booth renter*. These terms simply mean that the salon professional is independent and not an employee of the salon or organization. They are used interchangeably to be inclusive of all independent beauty-professional types.

reviewers

The author and publisher wish to acknowledge the following professionals for their support in the development of the manuscript by providing recommendations through their personal expertise as independent stylists.

Crystal A. Rivenbark
Hair Architecture, LLC, Midlothian, VA

Dawn Medina
Medina Hair Design, Fort Wayne, IN

Heather Murdoch
Guru Salon and Spa, Falmouth, ME

Janet Prediletto
Janet Prediletto Professional Aesthetics LTD., Saratoga Springs, NY

Linda Garcia
Independent Stylist, Rio Rancho, NM

Olivia Hill
The Hair Junkies Salon Studio, Allen, TX

Sandy LeClear
Trend Setters, Fort Wayne, IN

Tyler Fedigan
Saratoga Springs Massage Therapy, LLC, Corinth, NY

chapter 1

The Basics of Running a Booth-Rental Operation

CHAPTER OUTLINE

1.1 How to Decide Whether Booth Rental Is Right for You

1.2 Your Responsibilities as a Booth Renter

1.3 Summary

1.4 Top Takeaways: The Basics of Running a Booth-Rental Operation

Career Profile

Sara Lovell

Sara Lovell has been an independent stylist since 2007. After four years of experience working in a salon, she decided that she needed more flexibility to pursue on-location work at photo shoots and weddings. Although she knew that it would be risky, Sara decided to take the leap to becoming an independent stylist to further her career in the long run. Since making the transition, Sara has learned that her goals are within reach and that the potential for an independent stylist in this industry is endless.

Was there a particular reason for your decision to become an independent stylist?

I watched other stylists in the salon where I worked making transitions from being employees earning commissions and working fulltime to becoming independent stylists who seemed to make more money and have more time with their families and friends. I also wanted time to pursue other avenues of the industry.

Was the experience what you had initially anticipated?

Initially, I was very hesitant about making the change, but when I finally did, I soon learned that I had made the right decision. Thanks to the advice of the other seasoned booth renters and my sales representatives, I quickly came up with an inventory system and a schedule that worked for me to organize my chemicals, products, and supplies.

What challenges do you face now as a booth renter?

Being a successful booth renter definitely has its challenges. I used to worry about no longer having the stability of a consistent paycheck every week. However, I made sure that I had a good idea of how much business I had been generating as an employee so that when I made the switch to working independently, I always knew what my goals needed to be each week to be able to cover my bills, as well as my new expenses and the costs of running my business. Even now, when I feel good about my volume of business, there are always slow times. It is important for me to be prepared in case I have a slow week or become ill and miss a few days of work.

What negative experience taught you the lesson that you learned from the most?

I had become close friends with the owner of the salon where I previously worked, and we never felt it necessary to have a formal written contract of our casual business arrangement. By the time I chose to leave the salon, however, we had begun to argue. We had difficulty finding common ground about my decision to leave the salon because neither of us had taken the time to discuss and agree to any terms beforehand. Tension between us grew, and it was not long before my clients began to sense the tension during their appointments with me. After hearing about the first complaint from a client, I immediately realized that I could lose clients because of the situation, which was distracting me from focusing on providing my clients with the best possible experiences. From that point on, I promised myself that I would always be sure to remain completely professional while working with clients, even on difficult days. I also resolved that when I set up my next chair-rental agreement, I would first get everything in writing.

(Continued)

> **Do you have any advice or tips to share with future booth renters about choosing whether booth rental is right for them?**
> Yes. Ask for advice about whether or not booth renting would be right for you, even if it bruises your ego, because it is always a good idea to have outside opinions about how you run your business. Do your research, and be sure that you are financially stable. There are no guarantees for success, and everyone needs to be prepared for the unexpected. Working with a financial guide and a business consultant not only has shown me what my business is capable of achieving, but also has given me the tools and a plan to achieve my future goals. My dream of becoming a traveling educator and platform artist, and relocating to California, is closer than I had imagined.

Making a career shift as a booth renter is a huge decision for many seasoned cosmetologists and recent beauty-school graduates (Figure 1–1). The transition from employee to booth renter may seem to be an easy choice when you first think about it, especially if you are currently serving customers in a salon environment; however, what you might think and reality often can be two entirely different things. You might think to yourself, "How hard could it be working as a booth renter?" It is not what you do as an independent beauty professional that is difficult, however; it is running all the aspects of a business that can be tricky.

When you decide to operate as a booth renter, you actually sign up to be the owner of your own business. Some professionals are ready for business ownership, and some are not. The chapters to follow will help you decide whether becoming an independent stylist is the right career choice for you.

As you explore the basics of booth renting, it is important to first understand the definition of **booth renter**. A *booth renter*, as defined by the Internal Revenue Service (IRS), is "someone who leases space from an existing business and operates their own business as an independent contractor. A booth renter or independent contractor is responsible for his/her own record-keeping and timely filing of returns and payments of taxes related to their business."[1]

Operating a successful booth-rental business requires skill, knowledge, tenacity, and hard work. This business model can be very profitable and rewarding if the business is set up and operated correctly from the start. In this chapter, you will explore the basics of running a booth-rental business. As defined by the IRS, the terms *booth renter* and *independent contractor* are used synonymously. Because this text is designed specifically for those who lease space from existing businesses and operate their businesses independently, for the sake of clarity, the terms *booth renter* and *independent contractor* will also be used synonymously for the remainder of this text.

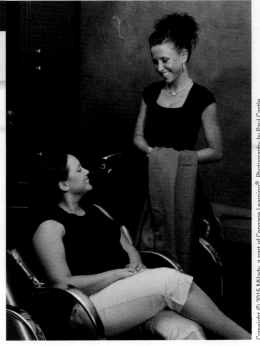

FIGURE 1–1

Starting Your Independent Beauty Business.

1.1 How To Decide Whether Booth Rental Is Right for You

Many beauty-industry professionals choose booth renting as a career choice for several reasons, including the following:

- They feel that they have outgrown their current salon environments but are not quite ready for salon ownership.
- They feel that they have reached certain levels of productivity and independence in their salon environments, and they desire an opportunity to control their income potential above and beyond what is currently offered in their places of employment.
- They want to establish their own work hours, schedule their own appointments, and be their own bosses.

The most common reason for salon professionals making the transition to booth-rental operations is money. These professionals know that they have excellent client retention and proven salary histories and that it is time for them to start reaping the benefits of keeping more of their hard-earned money in their own pockets. Although money is a strong motivator for making this change, a salon professional should carefully consider all the aspects of owning and operating his or her own business before making a crucial career change. While thinking about whether or not you are ready for booth rental, ask yourself the following questions:

- Am I an organized and independent worker?
- Do I enjoy keeping track of details about my clients, doing paperwork, and building up my clientele base?

HERE'S A TIP

According to the Professional Beauty Association report entitled "Economic Snapshot of the Salon and Spa Industry," of the 758,000 hairdressers, hair stylists, and cosmetologists working in the United States in 2011, 34 percent (or 260,000) were self-employed.[2]

- How many clients do I service in a week?
- What is my client-retention rate?
- How much is my monthly gross income?
- How much money am I generating in tips on a monthly basis?
- What is the average cost of hair products I use on a monthly basis? (Separate these products by category, for example, haircolor, shampoo, and conditioner.)
- What are my personal expenses?
- How much money can I afford, and how much am I willing to pay for booth rental each week?
- Will I need an assistant?
- Does your income allow for an assistant?
- How much will an individual medical plan cost me?
- Who will handle my accounting and record keeping?
- How will I handle the cleaning of my towels?
- What is my current marketing plan for obtaining new clients?
- How much will liability insurance cost?
- What percent of sales are run through a credit-card processing machine? How much will the fees be?
- In what type of environment do I want to rent a booth? (E.g., what values, vision, or other aspects of a work environment do you desire?)
- How much money will I have to set aside annually for ongoing education (both professional and business related)?

Answering these basic questions is crucial to helping you determine your readiness to operate as a booth renter.

Once you have decided that you are ready to transition into booth renting, it is important to start making preparations by reviewing some basic business and personal expenses to determine the amount of revenue that you will need to generate monthly to operate as a booth renter. Finances will be covered in more detail in Chapter 4, "Your Money, Your Future," but before you start the process of preparing to transition to booth renting, it is imperative to know how much money you must make to cover both your booth-rental business and your personal expenses.

Complete the Booth-Renter Readiness and Personal Income and Expenses Worksheets in Exercises 1–1 and 1–2 to give you a closer insight into what you need to do to prepare for your journey into booth renting. **Note:** If you do not have all the data required to complete Exercises 1–1 and 1–2, you can gather this information and complete the exercise later.

EXERCISE 1–1 Booth-Renter Readiness Worksheet

Fill in the information below as accurately as possible. Take some time to research your answers if necessary.

Average number of clients served in one week	
Client-retention rate (percentage of new clients who return for services)	

CALCULATE INCOME

Income	Monthly
Gross Income	$
Tips	$
Bonuses	$
Miscellaneous	$
TOTAL INCOME:	$

CALCULATE EXPENSES

Expenses	Monthly
Rent (Booth-Rental Fee)	$
Supplies/Utilities:	
Backbar Products	$
Towel Service	$
Cleaning	$
Business Phone/Cell Phone	$

Expenses	Monthly
Insurance:	
Professional Liability Insurance	$
Disability Income Insurance	$
Marketing:	
Retail Inventory Cost	$
Other Marketing/Advertising Expenses	$
Internet	$
Website	$
Gifts:	
Donations	$
Contributions	$
Other:	
Bookkeeping	$
Credit-Card Terminal Fees	$
Continuing Education	$
Assistant Wages	$
Receptionist Wages	$
Taxes on Income (typically paid quarterly)	$
TOTAL EXPENSES:	$

EXERCISE 1–2 Personal Income and Expenses Worksheet

Fill in the information below as accurately as possible. Take some time to research your answers if necessary.

CALCULATE INCOME	
Income	Monthly
Wages/Salary	$
Tips	$
Bonuses	$
Child Support	$
Alimony	$
Miscellaneous	$
TOTAL INCOME:	$

CALCULATE EXPENSES	
Expenses	Monthly
Housing:	$
Rent/Mortgage	$
Home Owner's Assoc. Dues	$
Home Repairs/Improvements	$
Insurance: Home/Renter's	$
Transportation:	
Car Payment	$
Insurance	$
Gas/Oil	$
Repairs	$
Bus/Subway/Taxi Fare	$
Parking Fees	$

Expenses	Monthly
Utilities:	
Electric	$
Gas/Fuel	$
Water	$
Trash/Sewer	$
Cable	$
Internet	$
Home Phone	$
Cell Phone	$
Home Security	$
Food:	
Groceries	$
Restaurants	$
Coffee/ Soda/ Snacks	$
Debts:	
Credit Card	$
Credit Card	$
Credit Card	$
Student Loan(s)	$
Medical:	
Insurance	$
Doctor	$
Dentist	$
Optometrist	$
Medications/Prescriptions	$
Personal:	
Clothing	$
Dry Cleaning/Laundry	$
Cosmetics/Styling Products	$
Child Care/Babysitter	$

(Continued)

EXERCISE 1–2   (Continued)

Expenses	Monthly
School Tuition	$
School Supplies	$
Magazines/Books	$
Memberships (Gym/Clubs)	$
Other Spending Money	$
Entertainment:	
Movies	$
Vacation	$
Music	$
Sporting Activities	$
Savings:	
Savings Account	$
Emergency Fund	$
401K/Stocks/Bonds/IRA	$
Gifts:	
Donations	$
Contributions	$
Gifts (*e.g.*, Birthdays, Christmas)	$
Other Expenses:	
Pet Care	$
	$
	$
TOTAL EXPENSES:	$
NET MONTHLY INCOME (Total Monthly Income Less Total Monthly Expenses)	$

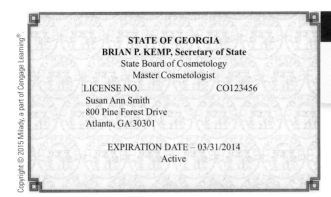

FIGURE 1–2

Renew Your License as Required by Your State or Local Government.

1.2 Your Responsibilities as a Booth Renter

As a booth renter, you are responsible for operating all facets of your independent beauty business. An employee-based salon carries the responsibility of providing their employees, the products, managing procedures, and overall business strategies needed to build a successful business. Operating independently as a booth renter requires that you obtain everything that you need to successfully run your business and serve your clients.

Let us take a look!

Making Sure That All Your Licenses Are Renewed and Up to Date

Each state has its own set of rules and standards for how often cosmetology licenses must be renewed. An example of a license is shown in Figure 1–2. Whatever your state mandates as a renewal process, you are required by law to renew your license(s) in order to continue practicing cosmetology in a commercial environment. Some local governments require that booth renters obtain a business license in order to run a booth-rental operation. Contact your local State Board of Cosmetology for license renewal procedures and costs. Contact your local state or county government to find out about business licensing requirements for running an independent booth-rental beauty operation in your area.

Setting Your Own Work Hours

As a booth renter, you are responsible for setting your own hours of operation (Figure 1–3). Before establishing your work hours, have a conversation with the salon owner about the salon's standard hours of operations. Having a dialog about the salon's business operations is important, because, for example, if a salon's standard hours of operation are Tuesday through Friday from 9 am to 6 pm, you may be required to set your business hours within this time frame. The salon

COACHING NOTE

If you are a recent beauty-school graduate, it is not recommended that you start your career as a booth renter. The business model of a booth-rental salon is not designed to build clientele for a booth renter, but rather to provide a place where experienced independent beauty professionals with steady clientele can service their clients. Therefore, if you do not have much experience and have few regular clients, it is not recommended that you choose booth renting as an option to start your career.

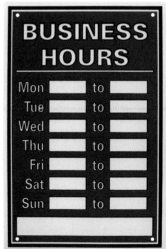

FIGURE 1–3

Set Your Own Hours.

owner is not required to open the salon outside of standard operating hours. However, if you and the salon owner agree to you having non-standard hours of operations, make sure that you include this into the written rental agreement. If you choose to work on an appointment-only schedule, which means that you are only available when you have an appointment, then the hours of operations listed on your business card should read, "By Appointment Only."

Scheduling Your Own Appointments

You are responsible for scheduling your own appointments with your clients so that these appointments coincide with your work schedule (Figure 1–4). There are several ways to schedule and keep track of your appointments. Many booth renters use a standard appointment book to keep track of their appointments. This type of book can be purchased from any supplier that carries beauty products; however, in today's world of technology, there also are software, Internet, and phone/tablet application programs that can be used to schedule appointments with your clients. The use of technology will help you better serve your clients and also speed up the process for you, saving you time and money. The choice of the appointment-scheduling tool is completely up to you. Whatever your choice, be sure to keep accurate records of all appointments for tax reporting purposes and keep all financial records for the time required by the IRS in case of an IRS audit. Refer to the IRS publication on recordkeeping guidelines, "Publication 583, Starting a Business and Keeping Records," on the website http://www.irs.gov. Overall, documents should be maintained at least for the period of time in which a claim could be brought based on, or related to, the documents (the relevant statute of limitations)[3].

FIGURE 1–4

Scheduling Your Own Appointments.

RESOURCES

Technology tools can be utilized to target a computer-savvy market. Some options may include, but are not limited to, the following:

- http://www.mychairapp.com
- http://www.styleseat.com
- http://www.clickbook.net
- http://www.schedulicity.com
- http://www.vagaro.com
- My Book: iPhone and iPad Application
- http://www.fullslate.com/
- http://www.saloniris.com

FYI

Even though it is required that a booth renter has his or her own phone line, in some instances, a salon will provide a business phone number for your clients to call and reach you. Some salons even employ front-desk receptionists to handle scheduling, and each independent stylist contributes to the salary of that receptionist. Be sure to ask the salon owners this question upfront when researching booth-rental locations. Additionally, rather than lumping all services into rent, it is not uncommon for an owner to offer additional services aside from rent for a fee, such as cleaning, electricity, phone/Internet, and co-op advertising.

Phone Service

As a booth renter, you are responsible for providing your own telephone contact number that is separate from the salon's phone number (Figure 1–5). The salon is not responsible for providing staff to take calls and appointments on your behalf. This means that clients will directly call your personal telephone number to schedule their appointments. If you are looking for a separate telephone service that is not connected to your personal cellular or home phone, it is recommended that you select one of the following options:

- Contact your preferred cellular provider and purchase a separate cellular phone in your business name so clients can call that phone to schedule appointments. You will be required to provide your business tax identification number to set up your account.
- Use a service such as Google Voice, by which you are given a separate phone number that rings into your current phone number of choice, which can be a cellular or landline phone number. When clients dial your Google Voice number, you will be notified with a distinctive message that the incoming call is coming from the Google-provided number.
- Contact your local phone company or private service provider and obtain a business call-forwarding telephone number. This type of service assigns you a unique number that is forwarded to your phone number of choice. The assigned business phone number will be listed in the local yellow pages along with your business name and will allow new and existing clients to easily find your business. This service is very efficient because it allows you to separate your personal calls from business calls and gives your business a unique telephone identity via caller identification.
- Install a separate business telephone line listed in your business name at your home. This phone number will be listed in the Yellow Pages and will allow new and existing clients to find your business under the Salon section of the Yellow Pages.
- Use a telephone answering service whereby the service provider will answer calls and make appointments on your behalf.

FIGURE 1–5

Provide Your Own Phone Service.

FYI

Did you know that obtaining a Google Voice number is free of charge? This service is great; it allows you to choose a phone number in the area code of your choice, which is perfect for booth renters. For more information on Google Voice, visit http://www.google.com/voice.

Setting Your Own Pricing for Services

The services that you decide to offer your clients and the prices that you charge for these services will be defined by you as a booth renter (Figure 1–6). One of the most common questions that booth renters have is, "How do I know what prices I should charge for my services?"

FIGURE 1–6
Set Your Own Prices.

STYLES BY DOTTI

<u>Haircuts</u>

Designer cuts for women	$40
Men's cut	$25
Children's cut	starting at $15
Formal updos	starting at $45

<u>Haircolor Services</u>

Virgin application, single-process	starting at $40
Color retouch	starting at $35
Double-process	starting at $55
Dimensional highlighting (full head)	$75
Dimensional highlighting (partial head)	$60

<u>Texture Services</u>

Customized perming*	starting at $80
Spiral perm*	starting at $100

Includes complimentary home-maintenance product.

Traditionally, the school of thought about this question has been to survey surrounding salons to find out about their pricing. However, solely focusing on what others charge for services is not a wise business practice. Think about it: Setting your prices based on what others charge would lead you to setting up your business based on others' experiences, clientele types, lifestyles, expenses, and profit goals. You would have no idea whether these other salons are running their businesses as you would like to run yours, nor whether the salons are successful. Therefore, their prices are not necessarily indicative of the level of service that you will be providing. Pricing should be established after considering a multitude of issues, such as supply and demand for your services (personal productivity), your expertise (skills and education), service experiences, location, and, most importantly, your personal expenses and goals for success. How to set your prices will be discussed in more depth in Chapter 7, "The Day-to-Day Details."

Product Supplies

Unless addressed differently in your contract, you are solely responsible for selecting and purchasing the brand(s) of products that you will use to perform services and sell at retail prices to your clients (Figure 1–7). Since you are independent, it is imperative that you keep an accurate inventory count of your backbar and retail products. Making sure that you have the products you need to serve your clients is essential. It is unprofessional to borrow products from other

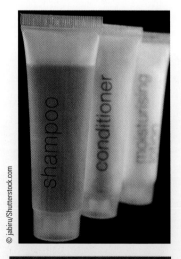

FIGURE 1–7
Provide Your Own Products.

salon professionals or to leave the salon to get the products that you need when a client arrives for an appointment. Remember, you want to maximize each client's experience with you, so it is important that you have retail products on hand to recommend to your clients at the end of their visit. You have only a limited space in which to keep your inventory, so make sure that you maintain a stock of your most popular backbar and retail products.

Operator Tools

All booth renters are responsible for providing the work tools that they need. For example, a hairdresser needs such tools as a blow dryer, hot irons, combs, brushes, razors, shears, clippers, shampoo capes, chemical brushes, gloves, towels, and other tools required to perform hairdressing services that are not part of the standard structural equipment that is provided by the salon (Figure 1–8). A salon is not responsible for providing booth renters with the tools they need to perform hair, nail, massage, or skincare services.

FIGURE 1–8

Supply Your Own Tools.

Collecting Money for Services Rendered to Clients

As a booth renter, you are responsible for collecting all money for services rendered to your clients (Figure 1–9). According to IRS rules and regulations, a salon cannot accept any money from clients whose services have been provided by a booth renter.[4] You might wonder, "Where should I put this money?" Many booth renters put money from clients in their aprons, purses, or pockets, but it is highly recommended that you purchase a lockable bank bag to store all of your cash and credit-card receipts.

FIGURE 1–9

Collect Your Own Money from Clients.

Professional Liability Insurance

Performing services for customers creates liability issues. There are many risks, and you can be held liable for any injuries to a client that occurred while being serviced in your chair. In order to protect yourself from these risks, it is important that you obtain professional liability insurance (Figure 1–10). **Professional liability insurance,**

FIGURE 1–10
Protect Yourself from Client Injuries.

also known as **Professional indemnity insurance**, is a specific type of insurance designed for independent contractors, such as self-employed beauty professionals. This type of insurance gives you protection in the event that you make a professional mistake, such as causing chemical burns to a client's scalp, causing a client's hair to fall out, burning a client with a hot iron, severely cutting a client's ear, or causing any other injury that a client sustains while you are performing services. Any injuries that a client may incur from you that could lead to that client obtaining legal counsel and pursuing the matter in court will be covered by your professional liability insurance policy. The amount of coverage that you will need depends on where you rent your booth and the type of services that you provide.

Disability Income Insurance

Once you become self-employed and this is the sole source of your income, protecting yourself in every aspect of your business is a priority. Self-employed people usually work very hard and do not always think about how they would provide for themselves if they were to contract some type of illness or sustain injury that could prevent them from working temporarily. This is why you need to protect your source of income by having disability income insurance (Figure 1–11).

FIGURE 1–11
Protect Your Income in Case of Disability.

Disability income insurance is insurance that will pay you a percentage of your income in the event that you become disabled or unable to work due to injury, illness, or any other incident that prevents you from performing your normal work activities. This type of insurance includes paid sick leave, short-term disability benefits, and long-term disability benefits.[5] In order to receive disability income insurance, you will be required to prove your earned income previous to the incident by producing an IRS W-2 wage form or annual tax return. Disability income insurance is important to have and should be included in your estimated monthly business expenses.

FIGURE 1–12

Pay Your Own Self-Employment Taxes.

Self-Employment Taxes

As a booth renter, you are responsible for paying your own self-employment taxes (Figure 1–12). The **self-employment tax** is how an independent contractor pays Social Security and Medicare payroll taxes. In an employee-based salon, the salon owner/employer and each salon employee split the cost of these payroll taxes. A booth renter is responsible for paying the total cost of this tax. It is important, therefore, for a booth renter to hire an accountant, partly because it may be recommended that you make quarterly estimated tax payments during the calendar year to ensure that your tax obligations are covered. A salon does not withhold taxes from a booth renter's pay because the salon does not handle any payments that booth renters receive from customers, as booth renters are not considered to be employees.

Estimated tax "is the method used to pay tax on income that is not subject to withholding,"[6] such as income received from clients for services rendered or earnings received from outside sources for services rendered by independent contractors. According to the IRS Publication *Tax Tips for the Cosmetology & Barber Industry*, "Estimated tax payments are made each quarter using Form 1040-ES, Estimated Tax for Individuals."[7] For more information or further instructions, contact your accounting tax professional.

> **HERE'S A TIP**
>
> There are many associations within the beauty and wellness industry that offer discounts on services and/or group rates for such things as health insurance, legal services, and liability insurance. Some associations to find information about include the following:
> http://www.insuringstyles.com
> http://www.probeauty.org
> http://www.ascpskincare.com (for skincare professionals)
> http://www.amtamassage.org (for massage practitioners)

FIGURE 1–13

Keeping Your Books Up To Date Is Important.

RESOURCES

You can organize receipts using software that scans, organizes, and stores your receipts. One type of software that handles paper-receipt organization is the NeatDesk Scanner. For more information on this system, visit http://www.neat.com.

Bookkeeping

You are responsible for keeping a record of all of your income and expenses. Hiring an accountant to do your bookkeeping is one way to keep track of your income and expenses, but there are also many software programs and applications that are designed to help you do your bookkeeping yourself (Figure 1–13). A word of advice: There will be many times when you purchase items and forget to record the transactions of these purchases, and sometimes your receipts can get lost or worn out. One of the best ways to separate spending for your business from your personal spending is to obtain a separate credit card that you will use for all of your business purchases. By using a separate business credit card, it is easier for you or your accounting professional to keep track of your spending and to maintain accurate records. If you spend cash on purchases, remember to always keep and organize your receipts. Cash purchases can only be tracked with receipts. If you feel that you might become overwhelmed by all the paper and organization that receipts cause, consider using software programs with scanners that allow you to scan receipts so that you can maintain a paperless system of tracking your receipts. (See Exhibit B: Sample Expense Tracker, at the end of this chapter.)

Retirement Plan

Saving for retirement is an important expense for a booth renter. It is recommended that you set a goal of saving 10 percent of your gross income for retirement savings. More details about saving for retirement is provided in Chapter 4, "Your Money, Your Future."

Rent

As a booth renter, you are responsible for paying rental fees for the space that you rent. The rent in a booth-rental salon is usually collected weekly by the salon owner or manager. How often you pay and the method of payment accepted is usually determined by salon management. As a booth renter, you should be given a receipt after each payment for tax purposes. At the end of the year, you must issue Form 1099-MISC for business rent paid of more than $600 to any noncorporate landlords each year.[8] Your accounting professional will create and file Form 1099-MISC on your behalf. Online programs and services also are available that will file and create your Form 1099-MISC for a small fee.

Towel Service

In many salons, towel service is not included in rent. Therefore, you are responsible for providing and cleaning the towels that you use for salon services (Figure 1–14). In your quest for seeking out a place where you would like to rent, ask salon owners or landlords if there are laundry facilities available for cleaning towels. If no laundry facilities are available, you may also search for companies that provide towel-cleaning and delivery services to salon professionals. Also confirm with salon owners whether towel storage space is available. Many salons will provide storage spaces or cabinets in the shampoo area for booth renters to store their towels.

FIGURE 1–14

Provide Your Own Towel Cleaning Service.

Work Station and Infection Control

It is the responsibility of the salon owner to keep the overall salon clean, but it is the responsibility of the booth renter to keep his or her work station and mirror clean of dirt, dust, hair, and debris (Figure 1–15). It is also the responsibility of the booth renter to keep all combs, brushes, towels, and implements cleaned and disinfected according to State Board rules and regulations.

Business Cards

It is the responsibility of the booth renter to provide his or her own business cards (Figure 1–16). The card style and design are the sole decisions of the booth renter. Your business card should have your business name as the main header and then state that your business is located at XYZ Salon, along with that address, your hours of operation, your contact phone number, and your website address. If the salon that you are thinking of renting from has a receptionist area, find out if the salon also has a designated space in the front area to display your business cards. This can be quite helpful to you in gaining potential clients who visit the salon while you are absent but want to take your business card to contact you for future services.

Marketing

As a booth renter, you are responsible for marketing your own business for new clients (Figure 1–17). Marketing includes providing and paying for your own website design

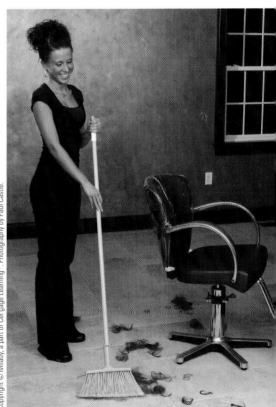

FIGURE 1–15

Keep Your Work Area Clean.

and Internet hosting. A more detailed discussion regarding marketing will be provided in Chapter 5, "Marketing Your Booth-Rental Business."

Hiring an Assistant

Many booth renters choose to hire assistants to help them. Before you rent a space, speak with the salon owner to learn about the policy for salon assistants. This is important because many booth renters who hire assistants use more than one chair in the salon and may be required to pay an additional fee for the use of an extra chair. If you hire an assistant, this assistant will be your employee. You will be required to check with your State Board of Cosmetology to find out what licenses are required, if any, for your assistant to support you in performing services for your clients. All payments, such as the assistant's wages, are your responsibility. As a business owner who has one or more employees, you must follow all wage requirements of your local

FIGURE 1–16

Provide and Select Your Own Business Cards and Design.

FIGURE 1–17

Create Your Own Marketing Plan.

government, as well as those of the state and federal governments. If you decide to hire employees, it is best that you contact a local payroll company to help you with payroll issues.

1.3 Summary

As you have learned, starting a booth-rental business requires a lot of time, energy, and thought. The decision to transition into a booth-rental operation is a big step, and you must be sure to understand some booth-rental basics. Let us explore some top takeaways from this chapter.

1.4 Top Takeaways: The Basics of Running a Booth-Rental Operation

- **The Transition to Booth Renting.** During the decision-making process of transitioning into a booth-rental operation, make sure that you consider your current income and your current personal and business expenses. These are among the first things to take into account before choosing to rent a booth. You want to be certain that you are currently generating enough revenue to cover your home and work expenses.

- **Your Responsibilities.** All facets of running your booth-rental business are your sole responsibility. The salon owner from which you rent is not responsible for scheduling your appointments, providing work products or tools, managing your inventory, providing personal liability or health insurance, building your clientele, paying your taxes, or keeping track of your revenue. When you start a booth-rental business, you are creating your own small business and you are in charge of all operations. Refer to the checklist provided at the end of this chapter for a complete list of responsibilities.

- **Insurance Coverage.** Since you are the sole revenue generator in your business, it is important that you protect yourself by obtaining health, liability, and disability insurance. Health insurance is necessary to help maintain your overall health, liability insurance will protect you in the event that a client is injured while you are providing services, and disability insurance will pay you a percentage of your earnings in the event that you become disabled.

> **HERE'S A TIP**
>
> Many booth renters receive cash on a daily basis and are tempted to spend it on lunch and other personal items during the day. Booth renters are required by law to report all income. Therefore, in order to keep your finances clean and simple, pay yourself each week, after you have made a deposit. Record your tips earned each day, deposit at least the amount necessary to cover taxes on those tips, and feel free to use the remainder for incidentals. Giving yourself a weekly wage allowance will help ensure that you have accurate income reporting and daily wage calculations.

- **Hiring an Assistant.** The most successful booth renters provide services for clients from the time they open until the time they close. In order to efficiently serve their clients, many booth renters do hire assistants.

---

[1] U.S. Department of the Treasury. (2011). *Tax tips for the cosmetology & barber industry*. Washington, DC: Internal Revenue Service. Retrieved on July 7, 2013, from http://www.irs.gov/.

[2] Professional Beauty Association. (2011). *Economic snapshot of the salon and spa industry*. Retrieved on July 5, 2013, from http://www.nysbsa.org/pdf/News_2011EconomicSnapshotoftheSalonIndustry.pdf.

[3] U.S. Department of the Treasury. (2011). *Starting a business and keeping records*. Washington, DC: Internal Revenue Service. Retrieved on July 7, 2013, from http://www.irs.gov/pub/irs-pdf/p583.pdf.

[4] *Ibid.*

[5] U.S. Bureau of Labor Statistics. (2008). *BLS information: Glossary*. Retrieved on July 7, 2013, from http://www.bls.gov/bls/glossary.htm/.

[6] U.S. Department of the Treasury. (2011). *Tax tips for the cosmetology & barber industry*. Washington, DC: Internal Revenue Service. Retrieved on July 7, 2013, from http://www.irs.gov/.

[7] *Ibid.*

[8] *Ibid.*

Chapter 1 Quiz: The Basics of Running a Booth-Rental Operation

This chapter has helped you to identify some basic steps needed to start your independent beauty business. Answer the following questions to review what you have learned.

1. You know that you are ready to operate a booth-rental business if you
   a. serve many clients.
   b. make over $500 a week in services and sales.
   c. have your business and personal finances in order.
   d. all of the above.

2. Booth renting is the perfect business model for students just graduating from beauty school.
   a. True
   b. False

3. List five responsibilities that you would have as a booth renter.

   1. _____

   2. _____

   3. _____

   4. _____

   5. _____

4. After a client has received your services, the client should pay the _____.
   a. front desk receptionist
   b. independent stylist
   c. salon owner
   d. none of the above

5. The person responsible for keeping your independent booth station clean is the_____.
   a. salon owner
   b. salon assistant
   c. independent stylist
   d. all of the above

6. The type of insurance that pays a claim when a client is injured while a booth renter is performing services for that client is_____.
   a. professional liability insurance
   b. health insurance
   c. general liability insurance
   d. occupational safety and medical insurance

7. An independent contractor should have _____ insurance, which will pay a claim if the independent contractor becomes disabled and cannot physically work for a period of time.
   a. dental
   b. disability
   c. health
   d. workers' compensation

8. The person responsible for paying employment taxes for a booth renter is the salon owner.
   a. True
   b. False

9. The salon owner is responsible for marketing and helping a booth renter build his or her clientele.
   a. True
   b. False

10. General liability insurance will protect a booth renter from any lawsuits, such as one that arises if a client has a serious allergic reaction from a chemical process.
    a. True
    b. False

11. Professional liability insurance will protect a booth renter in the event that someone is injured at the booth renter's place of business.
    a. True
    b. False

EXHIBIT A   The Basics of Running a Booth-Rental Operation

Use the checklist below to confirm that you have all the basics needed to run your own booth-rental operation.

BOOTH-RENTER RESPONSIBILITIES

- [ ] Valid/Current cosmetology license
- [ ] Current business license (if required)
- [ ] Appointment book/salon computer software scheduler or cellular phone application
- [ ] Phone service/Receptionist
- [ ] Service pricing list
- [ ] Back bar products inventory
- [ ] Retail inventory
- [ ] Operator tools purchased
- [ ] Method(s) to collect payments for services rendered
- [ ] Bank bag for storing income and credit-card receipts
- [ ] Professional liability insurance
- [ ] Disability income insurance
- [ ] Payment of self-employment taxes
- [ ] Bookkeeping/Accounting professional for bookkeeping and tax payments
- [ ] Continuing professional and business education for yourself
- [ ] Towel services
- [ ] Cleaning plan
- [ ] Business cards
- [ ] Website
- [ ] Marketing plan
- [ ] Assistant(s)

EXHIBIT B Business-Expense Tracker

Date	Check or Credit Card Number	Payee	Advertising	Auto Mileage	Insurance	Loan Interest	Legal and Professional Fees
			$		$	$	$
			$		$	$	$
			$		$	$	$
			$		$	$	$
			$		$	$	$
			$		$	$	$
			$		$	$	$
			$		$	$	$
			$		$	$	$
			$		$	$	$
			$		$	$	$
			$		$	$	$
			$		$	$	$
			$		$	$	$
			$		$	$	$
			$		$	$	$
			$		$	$	$
			$		$	$	$
			$		$	$	$
			$		$	$	$
			$		$	$	$
			$		$	$	$
			$		$	$	$
			$		$	$	$
			$		$	$	$
			$		$	$	$
			$		$	$	$
			$		$	$	$
			$		$	$	$
			$		$	$	$
Subtotal			$		$	$	$

EXHIBIT B Business-Expense Tracker (Continued)

Office Expenses	Supplies	Repairs and Maintenance	Wages	Taxes and Licenses	Travel	Meals and Entertainment	Utilities & Telephone
$	$	$	$	$	$	$	$
$	$	$	$	$	$	$	$
$	$	$	$	$	$	$	$
$	$	$	$	$	$	$	$
$	$	$	$	$	$	$	$
$	$	$	$	$	$	$	$
$	$	$	$	$	$	$	$
$	$	$	$	$	$	$	$
$	$	$	$	$	$	$	$
$	$	$	$	$	$	$	$
$	$	$	$	$	$	$	$
$	$	$	$	$	$	$	$
$	$	$	$	$	$	$	$
$	$	$	$	$	$	$	$
$	$	$	$	$	$	$	$
$	$	$	$	$	$	$	$
$	$	$	$	$	$	$	$
$	$	$	$	$	$	$	$
$	$	$	$	$	$	$	$
$	$	$	$	$	$	$	$
$	$	$	$	$	$	$	$
$	$	$	$	$	$	$	$
$	$	$	$	$	$	$	$
$	$	$	$	$	$	$	$
$	$	$	$	$	$	$	$
$	$	$	$	$	$	$	$
$	$	$	$	$	$	$	$
$	$	$	$	$	$	$	$
$	$	$	$	$	$	$	$
$	$	$	$	$	$	$	$
$	$	$	$	$	$	$	$
$	$	$	$	$	$	$	$
$	$	$	$	$	$	$	$

notes

chapter 2

Getting Your Business Off the Ground and Running

CHAPTER OUTLINE

2.1 Choosing a Business Structure

2.2 Licenses

2.3 Finding the Ideal Location to Rent a Booth

2.4 Work and Sublease Agreements

2.5 Understanding Your Legal Rights as a Booth Renter

2.6 Summary

2.7 Top Takeaways: Getting Your Business Off the Ground and Running

Career Profile

Sandy LeClear

As a licensed cosmetologist, esthetician, and educator with more than 30 years of success in the professional beauty industry, Sandy LeClear has used her creative flair, technical skills, and humor to navigate through nearly all aspects of the beauty business—from working on commission at a small, independent salon to being an eight-state district manager of a national salon chain. Becoming self-employed in 1989 allowed Sandy the flexibility to serve on the Indiana State Board of Cosmetology, continue to grow as a highly successful skincare expert, raise two daughters, and become a popular and versatile instructor of beauty culture. Being an independent stylist has given her the freedom to work for clients of her particular demographics and yet still have time to dedicate to teaching and furthering her own education by attending trade shows and classes.

How did you select the business structure for your independent beauty business? Did you work with an accountant and business attorney to make this decision?

When the three owners of the high-end salon where I worked made the decision to transition from a commission entity to booth rental, I started my solo career, ironically, as part of a herd. The owners, who were also part of our styling team, sat down with the staff and said, "Here is how booth rental works and these are the papers that you will file." So, we all followed their lead. Since it seemed to have been the right decision, I became a sole proprietor.

What are your wishes, or what strategies do you use to look for a location in which to rent?

All my senses engage when I am looking for a new place to rent. Even before I walk in the door, I note what is nearby that my existing clients can relate to as a selling point for why they should follow me to a new location. Once inside a salon, I make myself aware of its operations, look for a warm gesture from the owner or manager, and carefully observe people and the services being performed. I expect to see some genuine interest and friendly smiles from the beauty professionals. If they care about where they work, they are going to be evaluating me, just as I am evaluating them. I look for something that says to me, "Join us!" not, "Go away!".

Where are you currently working? Does the location of this salon affect your business and type of clientele that you work with?

Currently, I work at Trend Setters Hair Design Group. Trend Setters is the largest of three independent salons in what can only be called an "entertainment" plaza, which also houses a pub, three restaurants with bars, a performance venue for national recording artists, a tattoo studio, a comedy club, a gambling casino, limousine rental, a banquet hall, and various small businesses. The plaza area is bordered by a huge apartment complex, student housing for a large university, and residential neighborhoods, so the Trend Setters salon where I work has a very mixed base of clients.

What is the best business advice that you could give to a potential booth renter?

In my 30-plus years of being an independent stylist, I have made some risky decisions about changing locations. I find that it helps to really know your clients and what they enjoy doing. I promote their interests in amenities to my clients as if I am a local tour guide. I know the hip coffee shop; an awesome consignment shop; the best bakery; a quick exit off the highway; where to go to work out or tan; and where to find the best manicurists, massage therapists, and bartenders; and even where to get the most delicious chicken wings that my clients simply must try. I call it my "Working Tour" of city salons.

In Chapter 1, you learned some basic responsibilities of a booth renter. Now it is time to discuss the necessary steps to get your booth-rental business off the ground and running. As with starting any type of new business, there will be things that you are required to complete in order to successfully launch your business. Starting a booth-rental business requires planning, setting up a business structure, and making sure that you have the proper credentials required by your local government to assure that your business is operating in compliance with the law. Depending on the type of business structure that you select, you will be required to complete different tasks. In this chapter, the procedures needed to properly get your booth-rental business off the ground and running will be explored.

2.1 Choosing a Business Structure

Choosing a business structure is one of the most important steps that you will take as a booth renter. How your business is classified and how your business pays taxes is solely dependent upon the business structure. Before selecting a business structure, it is imperative that you give your business a name. When thinking about a name for your business, try to create a name that is both memorable and reflects your brand. Once you have selected a name for your business, it is time to proceed with choosing a business structure.

RESOURCES

For more help on building a successful brand, check out Milady's online course, *Branding Strategies for the Salon and Spa*.

When deciding upon which business structure will be best for you to operate your business, you should consult with an accountant or business attorney to help you make this decision. These professionals will consider several factors before helping you choose the business structure that is best for your independent beauty business. This section will discuss the four primary business structures of for-profit businesses: sole proprietor, limited liability corporation, S corporation, and C corporation (Figure 2–1).

Sole Proprietor

A **sole proprietor** is a business with a single owner who pays personal income tax on the earnings received from the business.[1] This business structure has few governmental regulations; therefore, many booth renters favor this business type. A sole proprietorship does not require that you create a separate trade name for your business, because you can operate the business under your own name. You may, however, create a fictitious, or "doing-business-as" (DBA), trade name for your sole proprietorship. (An example of a sole proprietor's trade name is

FIGURE 2–1

Choose a Business Structure That Is Best for You

"Jane Smith, DBA Salon Hair Trends"). Contact your county or state government to find out the required steps and fees for registering your trade name.

If you choose to operate your business as a sole proprietor, you are personally liable for all debts and obligations of the business, such as business loans and credit debts (Table 2–1). This means that if your business is sued and you are operating as a sole proprietor,

TABLE 2–1

Advantages and Disadvantages of a Sole Proprietorship

PROS	CONS
Total control and power of all decision making for the business.	Personal liability for all debts of the business, including credit-card debts and any business loans incurred by the business; if the business has employees, any financial debts or expenses acquired by employees on behalf of the business are also the personal liability of the sole proprietor.
Sole decision to sell or transfer the business.	Total control of business decisions and responsibilities.
No corporate tax payments.	Not appealing to investors because shares cannot be sold.
Very few costs involved.	Obtaining business credit is difficult because it is dependent on the owner's personal credit history.

your personal assets are at risk. According to a *New York Times* article, "In this type of business, there are no specific business taxes paid by the company. The owner pays taxes on income from the business as part of his or her personal income tax payments."[2]

Limited Liability Corporation

A **limited liability company (LLC)**, as defined by the IRS, is "a business structure allowed by state statute." LLCs are popular because, similar to corporations, LLC owners have limited personal liability for the debts and actions of their LLCs.

Owners of LLCs are called members. Since most states do not restrict LLC ownership, members may include individuals, corporations, other LLCs, and/or foreign entities. There is no maximum number of members allowed. Most states also permit single-member LLCs, which are those having only one owner."[3]

If you want to form an LLC, contact an accountant or select a professional online service that will create your LLC, register your business name, create your articles of organization, and file this document with your state authority. For the formation of an LLC, a document called **articles of organization** is required by every state government. (A few states, however, have a different name for this document, such as "articles of formation.") Articles of organization include the company's name and address, the name(s) and address(es) of the LLC's owner(s) and, if any, all member(s), the name of the **registered agent** (a person who is appointed to receive and send legal documents on behalf of the business), and a brief summary describing the purpose of the business. There is a fee to file an LLC's articles of organization; this fee varies from state to state.

An LLC is one of the most popular business structures among independent beauty professionals. This is because it allows for single-member ownership and also provides the protection that a corporation has (Table 2–2).

> **HERE'S A TIP**
>
> Many booth renters start their businesses as sole proprietors; however, an accountant may not recommend this structure for you if you plan to expand your business and build business credit. This structure is often reverted to by default because a business structure is not chosen when the business is formed. If you start your business without choosing a structure, for tax purposes, you will automatically be classified as a sole proprietor. It is fine to operate as a sole proprietor, but as your business grows and its liabilities increase, you may want to protect your personal assets from your business liabilities by changing your business structure.

TABLE 2–2 Advantages and Disadvantages of an LLC

PROS	CONS
Allows for pass-through taxation, so earnings can be taxed only once.[4]	Limitations on transfer of ownership of the business.
No liability for owner(s) or members for the company's debts unless they were required to sign a personal guarantee statement.	Fees and paperwork are required.

S Corporation

The IRS defines an **S corporation** as "a corporation that elects to pass corporate income, losses, deductions, and credit through to their shareholders for federal tax purposes. Shareholders of S corporations report the flow-through of income and losses on their personal tax returns and are assessed tax at their individual income tax rates."[5] This allows S corporations to avoid double taxation on corporate income.

An S corporation is required to file articles of incorporation with the state government. **Articles of incorporation** are legal documents that confirm the creation of a corporation. This document includes the company's address, as well as stock and shareholder information.

One of the major benefits of incorporating your business is personal-liability protection (Table 2–3). If your business is properly set up as a corporation and you do not personally guarantee your corporation's obligations, then your corporation—not you—will be responsible for its own obligations. One of the major reasons why people incorporate their businesses is to protect the personal assets of the owner(s) and shareholders, or stockholders, of the corporation.

TABLE 2–3 Advantages and Disadvantages of an S Corporation

PROS	CONS
Avoids double taxation by allowing pass-through taxation. which means the company's profits are passed through to the personal tax returns of the owner(s) and shareholders, and taxes are paid at an individual tax rate; the business does not pay corporate taxes.	Fee requirement to set up
No responsibility for the corporation's obligations, which provides protection of personal assets in the event the company faces any legal judgments.	Heavy documentation required in addition to articles of incorporation and frequent documented shareholder meetings

C Corporation

A **C corporation** is taxed separately from its owner(s). In a C corporation structure, the corporation pays federal and state income taxes on the company's earnings. Once the earnings have been distributed to the shareholders as dividends, the shareholders' incomes are also taxed, which means that a C corporation's earnings are taxed twice. This is different from an S corporation, in which the earnings are only taxed once. Many major corporations and some small businesses with substantial earnings operate as C corporations for tax purposes. A C corporation is also required to file articles of incorporation with the state government (Table 2–4).

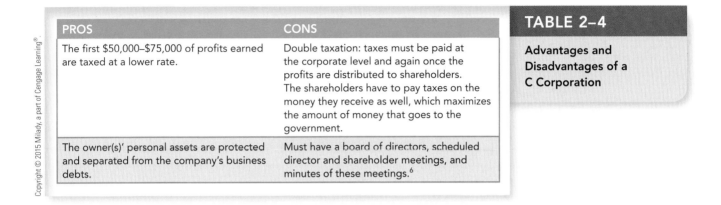

TABLE 2–4

Advantages and Disadvantages of a C Corporation

PROS	CONS
The first $50,000–$75,000 of profits earned are taxed at a lower rate.	Double taxation: taxes must be paid at the corporate level and again once the profits are distributed to shareholders. The shareholders have to pay taxes on the money they receive as well, which maximizes the amount of money that goes to the government.
The owner(s)' personal assets are protected and separated from the company's business debts.	Must have a board of directors, scheduled director and shareholder meetings, and minutes of these meetings.[6]

Examples of Corporation Structures

Deciding which business structure is best for your business can be an arduous task. The following scenarios are examples of the ways that different business structures work. These scenarios are provided to help with your corporate-structure selection process. For more information on selecting the best corporate structure for you business, you should contact an accountant.

Scenario 1

You are married, and your spouse's income is $60,000 a year. You and your spouse file joint income tax returns. For the first year of your business, your earned income was $50,000, but after you add up all the expenses and costs that you paid to keep the business running, your total expenses amount to $60,000 (this does not include taxes paid). So, for your first year of business, your actual profit is negative $10,000.00. During the fiscal year, taxes have been deducted from your spouse's income based on his annual salary of **$60,000.**

- Scenario 1, S Corporation: With an S corporation, the $10,000 loss in profit from your company would be carried over from your business tax return to your personal tax return. The $10,000 loss would then be deducted from your spouse's $60,000 income, which would change the adjusted gross taxable income before other personal deductions to $50,000. This is how a refund occurs: because your spouse had taxes deducted during the year based on his salary of $60,000, and now the household adjusted taxable income after your $10,000 business loss but before any personal deductions is **$50,000**.

- Scenario 1, C Corporation: With a C corporation, your $10,000 loss would not be transferred from the business taxes over to your personal income tax return but instead would remain with the company as a loss. This $10,000 loss would be carried over in the business to be deducted against future profits. In this scenario, your total household adjustable taxable income would stay at **$60,000**.

- Scenario 1, Sole Proprietor: As a sole proprietor, you would be required to fill out a Schedule C (Form 1040) to determine the net profit or loss of your business. The $10,000 loss would change your personal adjusted gross taxable income to **$50,000**.

Scenario 2

Your company is growing, and you are now in your third year of business. Your company grossed $200,000 during this fiscal year, and your expenses for the year were $70,000. You paid yourself an annual wage of $100,000, which you reported on your Form W-2. Your total company profit is now $30,000.

- Scenario 2, S Corporation: With an S corporation, the total profit for your company of $30,000 would be transferred over to your personal tax return, which would bring your total personal income to $130,000. With an S corporation, it does not matter how much you pay yourself; the total business profits would still be transferred over to your personal tax return. Using Scenario 1, in which your spouse earned $60,000, your total adjusted gross taxable income before personal deductions would be **$190,000**.

- Scenario 2, C Corporation: With a C corporation, the total profit for your company is $30,000 and your personal income is $100,000. Your company would personally issue you a Form W-2 for your wages of $100,000. Your company would then pay corporate taxes, which are generally at a lower rate than are personal tax rates on the $30,000 profit. You would pay personal taxes on your $100,000 income, and with your spouse's $60,000 income added, your total adjusted gross taxable income would be **$160,000**.

- Scenario 2, Sole Proprietor: As a sole proprietor, you would be required to fill out a Schedule C (Form 1040) to determine your net profit or loss. In this case, your profit is $130,000 and when your spouse's income is added, the total adjusted gross taxable income before any personal deductions is **$190,000**.

> **FYI**
>
> If you choose an LLC as a business structure, you can operate your LLC as an S corporation, as a C corporation, or as a single-member LLC, which is taxed as a sole proprietor.

Which Business Structure Is Best for You?

Complete Exercise 2–1 using your last five years of income and expenses to help you determine which business structure is best for you. If you have negative profits, your accountant would probably recommend that you operate as an S corporation or an LLC. If your business has large profits, your accountant may suggest that you operate as a C corporation.

EXERCISE 2–1 Business Structure Determination Chart

Year	Annual Income	Annual Expenses	Annual Profits
20__	$	$	$
20__	$	$	$
20__	$	$	$
20__	$	$	$
20__	$	$	$

Based on the past five years of your business history, which business structure is best for you?

_____

_____

_____

_____

2.2 Licenses

Starting a booth-rental establishment in the United States requires that you obtain the proper licenses and credentials needed to operate your business. You must have a valid cosmetology license obtained through the state in which your business will be established. Other required licenses are discussed below.

Cosmetology License

It is illegal for any person to practice cosmetology without a cosmetology license. You cannot rent a booth and operate your independent beauty business in any salon without this license. A salon that is offering chair rental will require that you have a valid cosmetology license in order to rent a space.

Federal Tax ID Number

An employer identification number (EIN), also known as a federal tax identification number, is used to identify a business entity. All businesses, regardless of structure, must obtain an EIN, except for a sole proprietor. If you choose to operate your business as a sole proprietor without employees, your social security number can be used as your taxpayer identification number. However, if you plan to hire employees for your sole proprietorship, you will be required to obtain an EIN. You may apply for an EIN in various ways.

Apply Online

You may apply for an EIN online by visiting http://www.irs.gov. Once you are on the website, go to the internal search field and enter "form ss-4 online." Click the search icon, and the link for the online EIN application will appear so that you can apply online. Once your online application is completed and submitted, you probably will be assigned your EIN number immediately, unless further information is needed. This is a free service offered by the IRS.

Apply by Phone

You may also obtain an EIN immediately by calling the IRS Business and Specialty Tax Line at (800) 829-4933. An IRS employee will take your information over the phone, assign your business an EIN, and provide this EIN to you as an authorized person over the telephone.

Note: International applicants must call (267) 941-1099 (not a toll-free number).

Apply by Mail or Fax

You can download form SS-4 from the IRS's website at http://www.irs.com. Once you have completed the form, you can mail it to Internal Revenue Service, Attn: EIN Operation, Cincinnati, OH 45999, or fax it to TIN: (859) 669-5760. (See Exhibit A: Application for an EIN.) For more information on obtaining an EIN, contact an accounting professional.

Business License

If you operate a business in a commercial space, it is usually required by your local government that you have a business license. A **business license** is a permit issued by a state, county, or federal government that allows an individual or company to conduct business within the government's jurisdiction. This license is the authorization to start a business. Although the salon where you rent a booth will have a business license, many states that allow booth renting require independent beauty professionals to obtain business licenses in order to operate as booth renters. For more information on business license requirements for booth renters in your area, contact your state or county government.

Sales Tax ID Number

If you are planning to sell retail products, you must obtain a sales tax Id number. A **sales tax ID number**, or **sales tax exemption certificate**, is a document issued by a state government that allows an individual or business to charge and collect local sales and use taxes on products and goods sold to customers. This certificate also allows you to purchase inventory from wholesale establishments and to file for tax-exempt status on the payment of goods that you plan to sell. Other names for a sales tax ID number include the following:

Reseller Permit	Sales Tax Exemption Certificate
Sales Tax Vendor ID Number	Certificate of Authority
Sales Tax Registration	State Tax ID Number
Reseller Tax ID	Reseller Certificate
Sales Tax Permit	

2.3 Finding the Ideal Location to Rent a Booth

Choosing where to rent a booth and house your establishment is vital to the success of your independent beauty business. Finding the right salon is important because the salon you select will play a part in your

FIGURE 2–2
Choose a Salon That Represents Your Independent Business Model

business model and brand representation. As shown in Figure 2–2, when in search for a place to rent a booth, there are five major things that you should consider.

1. **Existing Clientele**
   - Take a survey of where your existing clients are coming from. Do they work or live in the area of your existing salon location?
   - Do many of your clients travel to appointments with you on public transportation or by car?

   Answering these questions will help you decide where you should relocate or start your business.

2. **Location**
   - If the survey of your clients reveals that many of them travel on public transportation, find out if public transportation is available in the area where you are thinking of moving. If public transportation is a factor in your business success, you should select an area that offers this service.
   - Is there easy access to the location by car? This question is important for you to be able to continue serving clients who do not live in the area of the salon. If clients are pleased with your services, they will be willing to travel to a location that is easily accessible.
   - Is the salon in a high-traffic or low-traffic area? In a high-traffic area, a salon may get a lot of walk-ins, which will assist you in further building your clientele. A salon in a low-traffic area is not necessarily a bad thing, but it does mean that you will have to build your clientele through marketing efforts.
   - How far are you willing to drive from your home to work?
   - Is there enough parking at the salon location, especially during peak hours? The availability of parking spaces at a salon's location will also play a part in the success of your business. Make sure to select a salon location with a parking lot that can accommodate the clients and salon staff.

- Is there a fee for parking at the salon? If clients have to pay for parking, be mindful that this may have an impact on your existing and new clients.
- Is the salon located in a safe, low-crime area? Over the years, I heard clients say that they did not follow independent beauty professionals to new locations because they did not feel safe at those locations. Keep this in mind during your salon search.

3. **Salon Environment**
   - What type of clients do you currently serve? For example, are they mostly professional women and/or men, or are they younger, and perhaps more trendy or casual? The salon's environment and the type of clients whom you serve will play a huge role in your salon-selection process. You should select a salon whose environment is a match for your current clientele.
   - Are you looking for an upscale salon with a relaxing environment, or are you looking for a hip, happening salon with a lot of energy? Depending on the age range of your clientele and your personality, this will determine the type of salon environment you should choose.
   - Does the salon's décor and environment fit with the style of your existing clientele? Remember, when selecting a salon, make sure that the salon's environment is a good match for you and your clientele.
   - Is the salon clean and well-organized? A clean, organized environment should be a definite goal in choosing a salon.
   - What are the other booth renters, the owner, and their clientele like? Are they professional, friendly, and receptive to you? Getting along well with the other booth renters and the owner is imperative when selecting your next "home."

4. **Salon Floor Plan**
   - Would you prefer to work in a salon with an open floor plan, or would you rather service your clients in a private area, such as a salon suite? When considering what type of floor plan would suit you best, consider the types of services that you perform and decide whether a standard open floor plan will work for you or whether you need more of a private setting.

5. **Salon Amenities**
   - What type of amenities are you looking for in a salon? Does the salon have a laundry area and/or a break room?
   - Does the salon offer free Internet WiFi?
   - Based on the type of services that you offer, does the salon setup, including the workstation, retail, and shampoo areas, fit your needs?
   - Does the salon make accommodations for personal salon assistants?
   - Does the salon provide special services, such as complimentary coffee and water for customers?

Answering these important questions will help you find a salon that is a perfect match for you and your clientele. Your answers can

> **COACHING NOTE**
>
> When searching for a salon to rent a booth from, make sure that you randomly visit the salon several times. You should even schedule an appointment on a busy salon day to have a service performed. Visiting the salon during its peak hours will give you an insight into how the current booth renters get along with each other. There is no worse working environment than renting a booth in a salon where the other booth renters are rude and unfriendly. The current atmosphere of the salon will affect your business, so it is essential to check out the atmosphere ahead of time.

also prevent a mismatch between you and a salon. There are occasions when booth renters are not happy with the salon they choose, but they feel stuck because they have built a clientele at that salon location and do not want to pick up and move their entire clientele.

During your salon search, answer the questions in Exercise 2–2, Salon Search Comparison Chart, and you will avoid a lot of frustration later on.

EXERCISE 2–2 Salon Search Comparison Chart

	SALON NAME	SALON ADDRESS	SALON TELEPHONE
Salon A			
Salon B			
Salon C			

	SALON A	SALON B	SALON C
Does the salon have a current salon/business license posted?			
What are the weekly booth rental fees?			
Does the salon have an open floor plan, or salon suites?			
Is the salon's location served by public transportation?			
Is the salon located in a busy area to allow for walk-ins?			
Is there adequate parking?			
Is there a fee for parking?			

	SALON A	SALON B	SALON C
Is the salon close to your home?			
Does the salon environment meet your requirements?			
Is the salon clean and well-organized?			
Do the salon's hours fit your work schedule?			
Does the salon have a greeter in the reception area to welcome clients?			
Are the stations designed to accommodate retail products?			
Is there a laundry facility for washing towels?			
Is there proper towel storage in the shampoo area?			
Is there a salon policy that allows for assistants?			
Does the salon offer free WiFi?			
Does the salon have a break room?			
Does the salon offer a private room for consultation?			
Are the current booth renters friendly and welcoming?			

Open-Floor-Plan Salons Versus Salon Suites

In the savvy world of salons today, booth renters have two main options for the type of salon from which they can rent: open-floor-plan salons or salon suites. The differences between these two are explored below.

Open-Floor-Plan Salons

An **open-floor-plan salon** is designed as a large open space where beauty professionals share the common work space and the shampoo, color, and dryer areas (Figure 2–3). This floor plan does not allow for privacy; therefore, if you provide services that require privacy or if you simply prefer privacy, you may want to explore renting a salon suite.

Salon Suites

Salon suites provide independent beauty professionals with the freedom and flexibility of having their own private suites. Specifically, **salon suites** are a collection of mini-salons all under one roof, similar to a doctors' office building. Each suite is a separate room with a private, lockable entrance and is fully furnished with everything you need to run your own salon.

Salon suites are designed for salon owners who want to downsize, as well as for beauty and barber professionals presently working independently as booth renters in a conventional open-floor-plan salon and looking for an alternative (Figure 2–4). Salon suites offer independent beauty professionals the opportunity of salon ownership without expensive overhead costs.

FIGURE 2–3
Open Salon Floor Plan

FIGURE 2–4
Salon Suite Setup

Salon suites operate differently from traditional booth-rental salons and salons that hire independent contractors. First and foremost, the relationship between independent beauty professionals and a salon-suite owner is a tenant–landlord relationship. Therefore, these independent beauty professionals are regarded as being self-employed. Since there is no open-floor plan, independent beauty professionals are able to service their customers in a private, enclosed area.

Benefits of Operating in a Salon Suite:
- Having your own private, lockable entrance
- Creating your own personal atmosphere
- Service your clients in private
- Many salon suites are fully furnished

Commercial Space for Rent

Another option for an independent beauty professional is to rent a small commercial space that can be remodeled according to personal preferences. This has become another popular trend, as there are many available office parks that welcome independent salon professionals. When renting your own commercial space, it is imperative that you go through the terms of the commercial lease very carefully. It is especially important that you understand the part of the lease that states who is responsible for the repairs, upkeep, and heating and air conditioning (HVAC) system of the rented space. If you will be providing chemical and/or nail services, make sure to check the ventilation

> **HERE'S A TIP**
>
> When renting a commercial space, make sure that its heating and air-conditioning system will effectively cool and heat your space in the building according to your business needs. This is imperative. For example, if the air-conditioning system was designed to cool a building for business offices, you will probably need to have additional cooling installed for your space, because a salon business uses tools and equipment that produce heat, which will require additional cooling. You should also ensure that the building's hot-water heater has a large enough capacity to accommodate constant usage from you and all other tenants. If more cooling or a higher-capacity hot-water heater is needed for you to operate your business, you should have a discussion about who will pay for these additions with the landlord before the lease is signed.

system in the building, as proper ventilation is necessary. If you have difficulty understanding the terms of the lease, hire an attorney to review it so that he or she can provide you with a full understanding of the lease's terms and your responsibilities.

2.4 Work and Sublease Agreements

Once you have chosen a salon in which to rent a booth, make sure that the salon owner provides a work or sublease agreement for you to sign before you begin renting and servicing customers in that salon (Figure 2–5). Depending on the type of rental, a work agreement can be from 1 to 15 pages in length; however, the standard work agreement for booth rental is normally one or two pages. A work agreement will clearly state that the beauty professional is operating as an independent contractor and not as an employee of the salon. The work/rental agreement should state the following:

- The length of time of the agreement (e.g., six months or one year), often including a clause that allows for termination of the lease with a 30-day notice.
- The hours when the facility is open for you to service your clients.
- The amount of money that you will pay for rent and when the rent is due, along with any penalties for late payments.
- How the work station is to be used; for example, for providing hair and makeup services.
- Certain booth-renter responsibilities, such as workstation maintenance, the collection of money for salon services, and personal responsibility for filing and paying income taxes.
- A list of any additional or optional services and the fees associated with such services; for example, the use of a receptionist, a shared WiFi connection, maintenance, and cleaning.

See Exhibit B at the end of this chapter for an example of a booth-renter's work agreement.

A sublease is an agreement between a tenant who holds a master lease from its landlord and a third party tenant who

FIGURE 2–5

Make Sure You Understand Your Lease Agreement

is seeking to rent a part or whole section of space from the tenant who holds the master lease. The lease terms of a sublease may be set for part of or the entire lease term but cannot extend past the original terms of the master lease. In a sublease agreement, the tenant who holds the master lease is operating as a landlord and a lessor but he/she is the only person with responsibilities to the original landlord. Some master leases have clauses in them that don't allow for subleasing of building space, so make inquiry about this clause before signing this type of agreement to avoid sublease complications.

In most cases, open-floor-plan salons use a work agreement as a contract for booth rental, whereas salon suites, whose renters can sometimes completely operate as separate entities, will require a sublease. (See Exhibit C: Sublease Agreement Sample.)

2.5 Understanding Your Legal Rights as a Booth Renter

Before you sign any work agreement or lease at a salon, you should know what your legal rights are as a booth renter (Figure 2–6). When you rent a booth from a salon, you are considered to be self-employed. The relationship between a booth renter and the salon owner is a tenant–landlord relationship only. As a self-employed independent beauty professional, it is critical that you understand the difference

FIGURE 2–6

Know What Your Legal Rights Are as a Booth Renter

between an employee and an independent contractor. An **employee** is a person whose work is controlled by the terms of an employer. The employer controls the employee's work schedule, as well as what and how the work will be done. An employer is responsible for withholding income taxes from an employee's wages and is also responsible for paying unemployment taxes on the employee's earnings.

An **independent contractor** (booth renter) is a self-employed individual who provides services for an organization that has no control over what or how the services are performed.[7] The organization only has control over the outcome of the work. In other words, if you are a self-employed booth renter, a salon owner cannot do any of the following:

- Tell you what time to come to work to perform services.
- Tell you what amount to charge for your services.
- Tell you what products to use on your clients.
- Collect money from clients for whom you provide services.
- Tell you how to perform services.
- Tell you what products to recommend to your clients for maintenance at home.
- Keep you from selling retail products to your clients regarding how the products relate to the services that you provide.

Note: The items listed above *can* be added to the agreement signed by the owner.

A salon owner can include details in the contract about what services are permissible in the space that you are renting. For example, the salon owner can state in the contract or work agreement that the space that you rent can be used for hair services only, which means that you can only use the space you rent for these services only. Therefore, you cannot use the space to, for example, provide teeth-whitening services or any other services not related to hair. Before you sign any work agreements or contracts, be sure to discuss your business model with the salon owner so that everything is outlined in the contract.

Understanding your rights as a booth renter simply means you understand that you are a business within a business with a tenant–landlord relationship between you and the salon owner. If you still are not sure of your rights as a tenant or business owner, always feel free to contact an employment attorney.

> **HERE'S A TIP**
>
> Before you sign any work agreement or contract, it is important that you thoroughly understand your rights as they relate to the terms of the contract. Many booth-rental contracts have terminology that is similar to that of an employee-based salon. Contact a legal professional to review any contracts presented to you before signing them. Remember, once you sign a contract, you assume responsibility for everything that is included in the contract and is identified, by your signature, that you understand all terms to which you agree.

2.6 Summary

In conclusion, getting your booth-rental business off the ground and running requires much planning and preparation before you can actually start your business operations. Once you have finalized the basics of setting up your business, you can then begin to focus on creating a solid business model.

2.7 Top Takeaways: Getting Your Business Off the Ground and Running

- **Select a Business Structure.** When getting your business off the ground and running, it is important that you select a business structure. Selecting a business structure will determine how you pay your taxes.

- **Proper Licenses.** Operating a booth-rental business requires that you obtain the proper licenses necessary to run your business.

- **Ideal Salon Location.** Finding the right salon location to run your booth-rental business is the key to operating a successful booth-rental business.

- **Agreement Terms.** Once you have selected the location in which to operate your independent beauty business, make sure that you understand the terms of your work agreement or sublease.

- **Understand Your Rights.** It is important that you understand your legal rights as a booth renter. Operating a booth-rental establishment means that you are an independent contractor and not an employee; therefore, the relationship between you and the salon owner is a tenant–landlord relationship.

---

[1] Investopedia. (n.d.). Sole proprietorship: Educating the world about finance. Retrieved on July 14, 2013, from http://www.investopedia.com.

[2] *The New York Times*. (2007). "Advantages and disadvantages of sole proprietorships," All Business Breaking News, World News & Multimedia. Retrieved on July 14, 2013, from http://www.nytimes.com/allbusiness/AB4113314_primary.html.

[3] U.S. Department of the Treasury. (2012). "Limited liability company (LLC)." Washington, DC, Internal Revenue Service. Retrieved on July 14, from http://www.irs.gov/.

[4] Form-A-Corp Services. (n.d.). "LLC formation." Retrieved on July 14, 2013, from http://www.form-a-corp.com/llcs.

[5] U.S. Department of the Treasury. (2012.). "S Corporations." Washington, DC, Internal Revenue Service. Retrieved on July 14, 2013, from http://www.irs.gov.

[6] *Ibid*.

[7] U.S. Department of the Treasury. (2012). "Independent contractor defined." Washington, DC, Internal Revenue Service. Retrieved on July 14, 2013, from http://www.irs.gov.

Chapter 2 Quiz: Getting Your Business Off the Ground and Running

This chapter has helped you to identify what is needed to get your independent beauty business off the ground and running. Answer the following questions to review what you have learned.

1. Which one of the following credentials is necessary for independent stylists to practice cosmetology in the United States?
    a. Federal tax ID number
    b. Cosmetology license
    c. Business license
    d. Sales tax ID number
    e. All of the above

2. The purpose for obtaining a sales tax ID number is to_____.
    a. buy salon supplies
    b. sell retail products
    c. cover equipment costs
    d. none of the above

3. A business license is required to operate an independent beauty business.
    a. True
    b. False

4. Starting a booth-rental business requires that you choose a business structure.
    a. True
    b. False

5. List the four primary business structures used in for-profit businesses.
    1. _____
    2. _____
    3. _____
    4. _____

6. If you do not choose a business structure when starting your business, when taxes are due, your business will be classified as this business structure type: _____.

7. Which business structure pays taxes twice?
   a. Sole proprietor
   b. C corporation
   c. S corporation
   d. Limited liability corporation

8. When searching for a salon in which to rent a space, there are five things to consider. List these in the space below.
   1. _____
   2. _____
   3. _____
   4. _____
   5. _____

9. A salon suite is an enclosed floor plan in which an independent beauty professional is assigned his or her own private space with a lockable entrance.
   a. True
   b. False

10. The relationship between a booth renter and the salon owner is a tenant–landlord relationship.
    a. True
    b. False

11. An _____ is defined as someone whose work is controlled by the terms of employment.
    a. independent contractor
    b. employee

EXHIBIT A Application for EIN

Form SS-4
(Rev. January 2010)
Department of the Treasury
Internal Revenue Service

Application for Employer Identification Number
(For use by employers, corporations, partnerships, trusts, estates, churches, government agencies, Indian tribal entities, certain individuals, and others.)
▶ See separate instructions for each line.   ▶ Keep a copy for your records.

OMB No. 1545-0003

EIN

Type or print clearly.

1. Legal name of entity (or individual) for whom the EIN is being requested

2. Trade name of business (if different from name on line 1)

3. Executor, administrator, trustee, "care of" name

4a. Mailing address (room, apt., suite no. and street, or P.O. box)

4b. City, state, and ZIP code (if foreign, see instructions)

5a. Street address (if different) (Do not enter a P.O. box.)

5b. City, state, and ZIP code (if foreign, see instructions)

6. County and state where principal business is located

7a. Name of responsible party

7b. SSN, ITIN, or EIN

8a. Is this application for a limited liability company (LLC) (or a foreign equivalent)?   ☐ Yes   ☐ No

8b. If 8a is "Yes," enter the number of LLC members ▶

8c. If 8a is "Yes," was the LLC organized in the United States?   ☐ Yes   ☐ No

9a. **Type of entity** (check only one box). **Caution.** If 8a is "Yes," see the instructions for the correct box to check.
- ☐ Sole proprietor (SSN) _____
- ☐ Partnership
- ☐ Corporation (enter form number to be filed) ▶ _____
- ☐ Personal service corporation
- ☐ Church or church-controlled organization
- ☐ Other nonprofit organization (specify) ▶ _____
- ☐ Other (specify) ▶ _____
- ☐ Estate (SSN of decedent) _____
- ☐ Plan administrator (TIN) _____
- ☐ Trust (TIN of grantor) _____
- ☐ National Guard   ☐ State/local government
- ☐ Farmers' cooperative   ☐ Federal government/military
- ☐ REMIC   ☐ Indian tribal governments/enterprises

Group Exemption Number (GEN) if any ▶

9b. If a corporation, name the state or foreign country (if applicable) where incorporated

State | Foreign country

10. **Reason for applying** (check only one box)
- ☐ Started new business (specify type) ▶ _____
- ☐ Hired employees (Check the box and see line 13.)
- ☐ Compliance with IRS withholding regulations
- ☐ Other (specify) ▶
- ☐ Banking purpose (specify purpose) ▶ _____
- ☐ Changed type of organization (specify new type) ▶ _____
- ☐ Purchased going business
- ☐ Created a trust (specify type) ▶ _____
- ☐ Created a pension plan (specify type) ▶ _____

11. Date business started or acquired (month, day, year). See instructions.

12. Closing month of accounting year

13. Highest number of employees expected in the next 12 months (enter -0- if none).
If no employees expected, skip line 14.

| Agricultural | Household | Other |

14. If you expect your employment tax liability to be $1,000 or less in a full calendar year **and** want to file Form 944 annually instead of Forms 941 quarterly, check here. (Your employment tax liability generally will be $1,000 or less if you expect to pay $4,000 or less in total wages.) If you do not check this box, you must file Form 941 for every quarter.   ☐

15. First date wages or annuities were paid (month, day, year). **Note.** If applicant is a withholding agent, enter date income will first be paid to nonresident alien (month, day, year).

16. Check **one** box that best describes the principal activity of your business.
- ☐ Construction   ☐ Rental & leasing   ☐ Transportation & warehousing   ☐ Accommodation & food service   ☐ Wholesale-agent/broker
- ☐ Real estate   ☐ Manufacturing   ☐ Finance & insurance   ☐ Other (specify)   ☐ Wholesale-other   ☐ Retail
- ☐ Health care & social assistance

17. Indicate principal line of merchandise sold, specific construction work done, products produced, or services provided.

18. Has the applicant entity shown on line 1 ever applied for and received an EIN?   ☐ Yes   ☐ No
If "Yes," write previous EIN here ▶

Third Party Designee
Complete this section **only** if you want to authorize the named individual to receive the entity's EIN and answer questions about the completion of this form.

Designee's name | Designee's telephone number (include area code) ( )
Address and ZIP code | Designee's fax number (include area code) ( )

Under penalties of perjury, I declare that I have examined this application, and to the best of my knowledge and belief, it is true, correct, and complete.

Name and title (type or print clearly) ▶

Applicant's telephone number (include area code) ( )

Signature ▶   Date ▶

Applicant's fax number (include area code) ( )

For Privacy Act and Paperwork Reduction Act Notice, see separate instructions.   Cat. No. 16055N   Form **SS-4** (Rev. 1-2010)

EXHIBIT B Sample Booth-Renter's Work Agreement

SALON NAME HERE
WORK AGREEMENT (Booth Rental)

THIS AGREEMENT made and entered into on this date of _____ by and between **Salon Owner Business Name**, located in **Salon City and State**, hereinafter called **Salon Name**, and **Your Name,** a licensed cosmetologist, hereinafter called the Contractor,

WITNESSES THAT:

Contractor will render cosmetology services that include hair services, makeup applications, and eyebrow arching and waxing, but Contractor is not limited to any other services pertaining to cosmetology.

Contractor will perform service in accordance with the terms and conditions set forth below:

1. This agreement shall be in force for a period of 12 months. Contractor will work a probationary period of three months, during which this contract may be canceled without cause. This contract may be cancelled with at least 30 days' written notice from either party to the other. Contractor is responsible for one month's rent if a 30-day notice is not provided by Contractor. If any month's rent is not provided by Contractor within 10 days of its due date, Contractor will be subject to a court appearance for collection of the remainder of the balance. Upon cancellation of the contract, all rights and duties of the parties hereunder shall cease.

2. Contractor's services will be rendered to clients according to Contractor's own scheduling of appointments and Contractor will have full use of the premises during the hours of business operations only.

3. Contractor will rent a work station provided by **Salon Name** that will be used for cosmetology services only at the rate of $_____ per week payable every Saturday by 2:00 p.m. If payment is not received by 2:00 p.m. on every Saturday, a late fee of $25.00 will apply for each day that payment is late. If payment is not received within seven days, Contractor and his or her belongings will be removed from the premises. There will be an annual 5 percent increase on booth rental every July. A $_____ deposit is required on all stations, and this deposit will be refunded when Contractor terminates his or her contract if the station rented is in good condition. If the station is not in good condition, the deposit in the amount of $_____ will not be refunded but will be used to repair the station.

4. Contractor will be responsible for the appearance and/or daily maintenance surrounding the assigned work station. All materials needed to perform cosmetology services is the sole responsibility of the Contractor.

5. It is understood that Contractor is an independent contractor and that **Salon Name** has no control over the work to be performed hereunder or the results obtained. Contractor agrees, however, that said work shall be done in a manner that is acceptable by **Salon Name.** Contractor will hold **Salon Name** harmless from and against every and all claims arising in favor of any person, including Contractor, based on any substantiated personal injuries or personal injuries claimed to have arisen from the work performed by Contractor hereunder, whether directly or indirectly due to negligence by **Salon Name**, except in any instances of sole negligence by **Salon Name**.

(Continued)

EXHIBIT B (Continued)

6. Any and all monies given to Contractor from clients for services rendered belongs to Contractor, who does hereby accept full and exclusive liability for the payment of any and all taxes for unemployment insurance now or hereafter imposed by or under the laws of the United States, which are measured by the wages, salaries, or other remuneration paid to Contractor for work performed under the terms of this agreement. Contractor agrees to make payments of any and all such contributions and taxes or similar charges and to relieve **Salon Name** of any and all liability arising therefrom.

7. Contractor has received and agrees to abide by the rules and regulations set forth by **Salon Name**. If for any reason, Contractor does not uphold the standards of **Salon Name**, this agreement may be terminated without notice by **Salon Name**.

IN WITNESS WHEREOF, the parties hereto have executed this agreement on the day and year first above written.

SALON NAME   **INDEPENDENT CONTRACTOR (Booth Renter)**

By: _____   By: _____

EXHIBIT C Sublease Agreement Sample

This Commercial Sublease Agreement (Sublease) is made and effective on this date of _____ by and between **Salon Name** (Sublessor) and **Independent Contractor** (Sublessee).

By the terms of this lease agreement dated _____, **Landlord Company** (hereinafter referred to as Owner) leased certain space in the building located at **Address, City, County, State** to Sublessor commencing on the date of _____ and continuing through and including the date of _____. A redacted copy of this lease is attached as Exhibit A and made a part of this sublease, and that document is hereinafter referred to as the Master Lease.

1. **Demise of Premises and Terms**.

A. **Premises**

Sublessor makes available for lease a portion of the building designated as **Space Name** located within **Building Address, City and State** (the Subleased Premises).

Sublessor desires to lease the Subleased Premises to Sublessee, and Sublessee desires to lease from Sublessor the space in the building identified above, according to the rental agreement and based upon the covenants, conditions, and provisions herein set forth.

THEREFORE, in consideration of the mutual promises contained herein, consideration of other goods and valuables, it is agreed:

B. **Terms**

Sublessor hereby leases the Subleased Premises to Sublessee, and Sublessee hereby leases the same from Sublessor, for an initial term, beginning on the date of _____ and ending **on** _____. Sublessor shall make every effort to give Sublessee possession of the Subleased Premises as nearly as possible to the date of the beginning of the Sublease terms. If Sublessor is unable to timely allow Sublessee to take possession of the Subleased Premises, rent shall abate for the period of delay. Sublessee shall make no other claim against Sublessor for any such delay.

2. **Rental**

Sublessee shall pay to Sublessor during the Initial Term rental a rate of **$**_____ per week. Each installment payment shall be due in advance on each Saturday during the lease term to Sublessor, and this payment shall be placed in the drop box located inside the salon at **Salon Address** or at such other place designated by written notice from Sublessor or Sub lessee. Weekly rent payment received after 12:00 p.m. on the Monday following the Saturday due date shall be subject to a $50.00 late fee per each occurrence.

This is a 12-month Sublease with installments of **$**_____ per week.

Sublessee has provided a security deposit in the amount of **$**_____, which is in escrow with Sublessor.

(Continued)

EXHIBIT C (Continued)

3. Use

Notwithstanding the foregoing terms, Sublessee shall not use the Subleased Premises for the purposes of storing, manufacturing, or selling any explosives, flammables, or other inherently dangerous substances, chemicals, things, or devices. Sublessee covenants and agrees to use the Subleased Premises in accordance with the terms and conditions of the Master Lease (for hair and beauty services only) and further covenants not to commit any act that will result in a violation of the terms of the Master Lease.

4. Sublease and Assignment

Sublessee shall not, without the prior written consent of Sublessor and Owner, assign the terms of this Sublease, nor permit it to be assigned by operation of law or otherwise; nor shall Sublessee, without the prior written consent of Sublessor and Owner, sublet all or any part of the Subleased Premises or permit the Subleased Premises or any part of them to be used by others for hire.

5. Repairs

During the sublease term, Sublessee shall pay in full for all necessary repairs to the Subleased Premises and assigned equipment and fixtures. Repairs shall include such items as routine repairs of floors, walls, ceilings, chairs, cabinetry, shampoo bowls, hair traps, all plumbing attached to shampoo bowls that is not encased in the walls/sheetrock, and other parts of the Subleased Premises and equipment damaged or worn through normal occupancy, except for major mechanical systems, subject to the obligations of the parties otherwise set forth in this Sublease.

6. Alterations and Improvements

Sublessee, at Sublessee's expense, shall have the right, with Sublessor's prior consent, to decorate or enhance all or any part of the Subleased Premises as Sublessee deems desirable, provided that decorations and enhancements are made in a professional manner and with good-quality materials. Sublessee shall have the right, with Sublessor's prior consent, to place and install personal property, equipment, and other temporary installations in and upon the Subleased Premises and to fasten these to the premises provided that (1) there is no interference or disruption of services to other tenants or to the facility at-large and (2) there is no damage to the Subleased Premises and/or assigned equipment. All personal property, equipment, machinery, and temporary installations, whether acquired by Sublessee at the commencement of the Sublease term or placed or installed on the Subleased Premises by Sublessee thereafter, shall remain Sublessee's property, free and clear of any claim by Sublessor. Sublessee shall have the right to remove such property at any time during the term of this Sublease, provided that any and all damage to the Subleased Premises and/or assigned equipment and fixtures caused by such removal shall be repaired by Sublessee at Sublessee's expense.

7. Property Taxes

Sublessor shall pay, prior to delinquency, all personal property taxes with respect to Sublessor's personal property, if any, on the Subleased Premises. Sublessee shall be responsible for paying all personal property taxes with respect to Sublessee's personal property at the Subleased Premises.

EXHIBIT C (Continued)

8. Insurance

A. If the Subleased Premises or any other part of the building is damaged by fire or other casualty resulting from any act of negligence by Sublessee or any of Sublessee's agents, employees, or invitees, rent shall not be diminished or abated while such damages are under repair, and Sublessee shall be responsible for the costs of repair not covered by insurance.

B. Sublessor shall maintain fire and extended coverage insurance on the building and the Leased Premises in such amounts as Sublessor shall deem appropriate. Sublessee shall be responsible for expenses for fire and extended coverage insurance on all of its personal property, including removable trade fixtures, located in the Subleased Premises.

C. Sublessee and Sublessor shall, each at its own expense, maintain a policy or policies of comprehensive general liability insurance related to their respective activities in the building, with the premiums thereto to be fully paid on or before the due date, issued by and binding upon an insurance company approved by Sublessor. Such insurance shall provide minimum protection of not less than $300,000 combined single-limit coverage of bodily injury, property damage, or a combination thereof. Sublessor shall be listed as an additional insured on Sublessee's policy or policies of comprehensive general liability insurance, and Sublessee shall provide Sublessor with current Certificates of Insurance evidencing Sublessee's compliance with the terms stated in this Paragraph. Sublessor shall obtain the agreement of Sublessee's insurers to notify Sublessor that a policy is due to expire at least 10 days prior to such expiration. Sublessor shall not be required to maintain insurance against thefts within the Subleased Premises or the building. However, Sublessor will not be held responsible for replacement or replacement value of any items stolen from Subleased Premises or the building.

9. Utilities

Sub lessee shall pay all charges for telephone and other services and utilities (except water, gas, electricity) used by Sub lessee on the Subleased Premises during the term of this Lease unless otherwise expressly agreed in writing by Sub lessor. Sub lessee acknowledges that the Subleased Premises are designed to provide standard salon use of electrical facilities and standard office lighting. Sub lessee shall not use any equipment or devices that utilize excessive electrical energy or which may, in Sub lessor's reasonable opinion, overload the wiring or interfere with electrical services to other tenants. Microwaves, refrigerators, space heaters, and additional hooded dryers are not permitted in Sub lessee's mini-salon suite.

10. Signs

With Sublessor's prior consent, Sublessee shall have the right to place on the Subleased Premises, at locations designated by Sublessor, any signs that are permitted by applicable zoning ordinances and private restrictions. Sublessor may refuse consent to any proposed signage that Sublessor deems to be too large, deceptive, unattractive, or otherwise inconsistent with or inappropriate to the Subleased Premises or any other tenants/sublessees.

(Continued)

EXHIBIT C   (Continued)

11. Entry

Sublessor shall have the right to enter the Subleased Premises at reasonable hours to inspect the same, provided that Sublessor shall not thereby unreasonably interfere with the operation of Sublessee's business on the Subleased Premises.

12. Parking

During the term of this Sublease, Sublessee shall have the nonexclusive use in common with Sublessor, other tenants of the building, and their guests and invitees, of the nonreserved common parking areas, driveways, and footways, subject to rules and regulations for the use thereof as prescribed at any time by Sublessor. Sublessor reserves the right to designate parking areas on the building's premises or in reasonable proximity thereto, for Sublessee and Sublessee's agents and employees. Sublessee shall provide Sublessor with a list of the license-plate numbers of all cars owned by Sublessee, its agents, and employees.

13. Building Rules, Policies, and Procedures

Sublessee will comply with the rules of the building adopted and altered by Sublessor at any time and will cause all of its agents, employees, invitees, and visitors to also comply with such rules. All changes and additions to such rules will be sent by Sublessor to Sublessee in writing. The initial rules for the building are attached hereto as Exhibit B and incorporated herein for all purposes. Sublessee will comply with the policies and procedures set forth in Exhibit B within the specified timeframes.

14. Damage and Destruction

Subject to Section 8 A. above, if the Subleased Premises, any part thereof, or any appurtenance thereto is so damaged by fire, casualty, or structural defects that such parts or appurtenances cannot be used for Sublessee's purposes, then Sublessee shall have the right within 90 days following such damage to elect by written notice to Sublessor to terminate this Sublease effective on the date of such damage. In the event of minor damage to any part of the Subleased Premises, and if such damage does not render the Subleased Premises unusable for Sublessee's purposes, Sublessor shall promptly repair such damage at the cost of Sublessor. In making the repairs required in this Paragraph, Sublessor shall not be liable for any delays resulting from strikes, governmental restrictions, inability to obtain necessary materials or labor, or other matters that are beyond the reasonable control of Sublessor. Sublessee shall be relieved from paying rent and other charges during any portion of the Sublease term in which the Subleased Premises are inoperable or unfit for occupancy, or use, in whole or in part, for Sublessee's purposes. Rent payments and other charges paid in advance for any such time periods shall be credited on the next ensuing payments, if any; however, if no further payments are to be made, any such advance payments shall be refunded to Sublessee. The provisions of this Paragraph extend not only to the aforesaid matters, but also to any occurrence that is beyond Sublessee's reasonable control and that renders the Subleased Premises, or any appurtenance thereto, inoperable or unfit for occupancy or use, in whole or in part, for Sublessee's purposes.

EXHIBIT C (Continued)

15. Security Deposit

The security deposit shall be held by Sublessor without liability for interest and as security for the performance by Sublessee of Sublessee's covenants and obligations under this Sublease, with it being expressly understood that the security deposit shall not be considered an advance payment of rent or a measure of Sublessor's damages in case of default by Sublessee. Unless otherwise provided by mandatory nonwaivable laws or regulations, Sublessor may commingle the security deposit with Sublessor's other funds. Sublessor may, at any time, without prejudice to any other remedy, use the security deposit to the extent necessary to make good any arrearages of rent or to satisfy any other covenant or obligation of Sublessee hereunder. If the amount of arrearages of rent or any other covenant or obligation exceeds the available value of the security deposit, Sublessee remains personally responsible for any and all balances due to Sublessor. Following any such application of the security deposit, Sublessee shall pay to Sublessor on demand the amount so applied in order to restore the security deposit to its original amount. If Sublessee is not in default at the termination of this Sublease, the balance of the security deposit remaining shall be returned by Sublessor to Sublessee. If Sublessor transfers its interest in the Subleased Premises during the term of this Sublease, Sublessor may assign the security deposit to the transferee and thereafter shall have no further liability for the return of such security deposit.

16. Notice

Any notice required or permitted under this Sublease shall be deemed sufficiently given or served if sent by U.S. certified mail, with a return receipt requested, and addressed as follows:

If to Sublessor, to: **Business Name and Address**

_____

If to Sublessee, to: **Sublessee Name and Address**

_____

Sublessor and Sublessee shall each have the right at any time to change the place at which notice is to be given under this Paragraph by written notice thereof to the other party.

17. Governing Law

This Agreement shall be governed, construed, and interpreted by, through, and under the laws of the State of **State**.

IN WITNESS WHEREOF, the parties have executed this Sublease on the day and year first written above.

Signed, sealed, and delivered to Sublessor and Sublessee in the presence of:

_____   _____
Witness   **Salon Business Name** (Sublessor)

(Continued)

EXHIBIT C (Continued)

Independent Contractor (Sublessee)

_____

Sublessee's Social Security Number and Date of Birth

Sublessee's Address:

_____

_____

© DeShawn Bullard

chapter 3

Eliminate Detours and Distractions with a Solid Business Model

CHAPTER OUTLINE

3.1 Create a Vision and Mission Statement

3.2 Set Goals for Your Business

3.3 Identify Your Core Business Values

3.4 Create a Solid Business Plan

3.5 Growing the Business

3.6 S.W.O.T. Analysis

3.7 Summary

3.8 Top Takeaways: Creating a Solid Business Model

Career Profile

Edda Coscioni

Edda Coscioni has been an independent stylist since 2008. She found herself at a career standstill and knew she needed to take the risk to go out on her own to become an independent stylist. Though quick, the transition was not always smooth for her. After working hard to establish her independent business, she has accomplished much, including becoming the reigning champion of fashion hairstyling for the United States. She was also chosen to compete at the World Championships in 2012 as part of the United States Team.

Can you describe what it was like for you in the beginning of your career as an independent stylist?

I had not planned on becoming a booth renter. When I started, I really had no sense of what it would take to have my own business. The whole shift came very quickly. With that in mind, I started to immediately work on how to build my clientele. There were times I felt overwhelmed. There were so many things to do and learn. Knowing that I needed to work on building my business, I joined a business-support group and a sales-support group. They helped me set my goals and accomplish them. I also built a client-tracking system, and tracked down all my old clients, making sure they knew where to find me.

What are some of the biggest challenges as an independent stylist?

The biggest challenge I face is how to grow my business and get "the word out" about it. This requires going out and becoming a part of various Chambers of Commerce and other business groups, as well as having a great referral program for my clients. This is ongoing work. One other issue is that I am in constant negotiations with the owner of the salon from which I rent. Sometimes, this is frustrating, but I still love that I am an independent professional.

A S.W.O.T. analysis is a planning method used to identify the strengths, weaknesses, opportunities, and threats of your business. What would you say are the strengths and weaknesses of your business?

Our industry is amazing. There are so many ways you can continue to grow in our field. Furthermore, as an independent beauty professional, you are not only the stylist, you are the CEO and the sales and marketing staff too. Using S.W.O.T. analysis is a great way to both understand my strengths and build on them, and to know my weaknesses so that I am always learning and becoming better. For example, one of my strengths is that I came from corporate education before I went into cosmetology. I use this strength to teach my clients how to do their hair. I also write a blog for my clients, educating them and inspiring them to try new things. One of my weaknesses I am working on is financial projections and creating solid profit and loss (P&L) sheets. P&L sheets allow me to fully comprehend what I make each year, where my largest returns on investment are, and what my biggest costs are. Knowing this, I am becoming more strategic in how I spend my time.

What advice would you give to future independent stylists?

It is so important to have a clear vision and mission statement for you and your business. The more you can build a foundation, the more likely you will be successful. Without it, you might not be able to grow your business, or worse you might fail. The clearer you are in setting goals for your business, identifying core business values, creating a solid business plan, and growing the business, the more you will understand what you need to do to be successful, profitable, and satisfied in your career. If it feels overwhelming, get a business coach. If you cannot afford one, check out your local colleges or look on-line for organizations that give business coaching for free.

Starting a business is an exciting experience, but every business needs a solid business model. Many beauty industry professionals are familiar with creating a business plan to get started in business and to help give them direction. Many times, creating a business model is left undone, but it is very important to create one. You may be thinking, what is the difference between a business plan and a business model? The difference is that a **business plan** is a written description of your business as you see it today, and as you see it in the next five years (detailed by year). It outlines the marketing plan, the equipment you need to operate your business, how to build and retain clientele, the staff you need to run the business, and describes how financially stable the company should be.[1] A **business model**, on the other hand, defines how your business will make money. It has components similar to a business plan, but it goes further by providing expense details. It focuses on the products and services that generate the most revenue and shows where your business will make the greatest profit (Figure 3–1).[2]

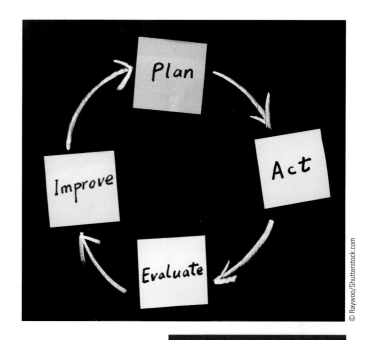

FIGURE 3–1

Create a Solid Business Model.

Creating a business model is crucial because it will define a clear roadmap as to how your business will make a profit. A business model focuses on the following key components:[3]

1. **Primary Partners**
   - Who are your primary business partners or suppliers?
2. **Primary Activities**
   - What is your primary business activity?
3. **Primary Resources**
   - What supplies or personnel are needed to provide the service?
4. **Value Proposition**
   - What value added services and products does your business offer?
5. **Customer Relationship**
   - How do you build customer loyalty?
6. **Target Market**
   - Who are your customers?
7. **Revenue Source**
   - What is your primary stream of revenue?

8. **Customer Reach Channel**
   - What channels will you use to reach new customers?
9. **Cost Structure**
   - What will it cost to provide the products or services?

Now it is time to define the components of your business model. Fill out the business model template in Exercise 3–1 to find out some facts about what it will take to turn a profit for your business.

EXERCISE 3–1 Business Model Template

PRIMARY PARTNERS Who are your suppliers? (Consider product, implements, towel service provider, business cards, etc.)	1) 2) 3) 4) 5) 6) 7)
PRIMARY ACTIVITIES List your primary business activities. What service or products will your business provide?	1) 2) 3) 4) 5)
PRIMARY RESOURCES What key resources are required to provide services? (Example: personnel, supplies, etc.)	1) 2) 3) 4) 5) 6) 7)

VALUE PROPOSITION What services does your business offer that have the most value to your customers?	1) 2) 3) 4)
CUSTOMER RELATIONSHIP How will you build relationships and customer loyalty to provide services to your target market?	1) 2) 3) 4) 5) 6) 7)
TARGET MARKET Who are your primary customers? (Demographic—age, median income, professional, student, male, female, etc.)	1)
REVENUE SOURCE List the ways your business plans to make money.	1) 2) 3) 4) 5)

(Continued)

EXERCISE 3–1 Business Model Template (*Continued*)

CUSTOMER REACH CHANNEL

What channels do you plan to use to reach your target market? (Example: direct mail campaign, billboard, social media, etc.)

1)
2)
3)
4)
5)
6)
7)

COST STRUCTURE

What expenses are necessary to provide your primary services? (Example: booth–rental fees, marketing, personnel, business services, etc.)

1)
2)
3)
4)
5)
6)
7)
8)
9)
10)
11)
12)
13)
14)
15)

3.1 Create a Mission and Vision Statement

Many people have dreams and visions of the type of business they want to own and how they want to run their business operations (Figure 3–2). The way you successfully turn your dream into a reality is to create a mission and vision statement for your business. A **mission statement** is a statement that describes who you are and what you do. Your mission statement should also briefly describe the what, why, and how of your business.

FIGURE 3–2

Create a Mission Statement for Your Business.

Example of a Mission Statement

At XYZ Salon, our service promise is to "delight" our guests. We are committed to creating one-of-a-kind hair-care experiences through our custom consultation, professionalism, and excellence in education.

A **vision statement** is a statement that describes your company's futuristic goals, its values, and its direction. It should state the dream you have for your business in the years to come and what it should look like when it gets there (Figure 3–3).

You deserve the most prosperous and successful business you could ever dream of. Continue through the chapter on goal setting and values creation. Later in this chapter you will write your own mission and vision statements.

FIGURE 3–3

Create a Vision Statement for Your Business.

Example of a Vision Statement

Example: XYZ Salon Service is a global service and retail company providing custom wig services, as well as specially formulated products and custom wigs designed to help the growing number of women experiencing hair thinning and hair loss.

3.2 Set Goals for Your Business

Once you have created your business model and your mission and vision statements, you will need to set some goals for your business to keep you on track as you work to achieve the vision of your company. When you set goals for your business, it is just like creating a roadmap for how you will fulfill your company's vision. Setting goals for your business requires two main components: setting short-term and long-term goals.

Short-term Goals

Short-term goals are those you set for your business within the first two years of starting it. Sometimes, your short-term goals will be immediate goals that you plan to initiate within the first three months of starting your business. When you create your short-term goals, keep in mind that it is a path to achieving your long-term goals. Short-term goals should be S.M.A.R.T. (specific, measurable, achievable, realistic, time specific) (Figure 3–4):

- **Specific:** Create specific, detailed objectives that will be implemented in the immediate future.
- **Measurable:** Determine what it will cost to implement your objectives. Detail each objective and attach a value to each objective to determine what it will cost.
- **Achievable:** Determine the who, when, and what you need to accomplish your objectives.
- **Realistic:** Take a good look at what you actually have to work with and create realistic goals you can truly accomplish.
- **Time Specific:** Create a timeline you will follow to help you complete your goals in a timely manner.

FIGURE 3–4
Your Short-term Goals Should be S.M.A.R.T.

Long-term Goals

Long-term goals are goals you plan to achieve for your business within three to five years. These goals are usually created with your company's mission and vision statements in mind (Figure 3–5). The short-term goals you set will be used as a platform for creating your long-term goals.

Your long-term goals should include the following four areas:

- **Service:** This particular goal should focus on customer service, satisfaction, and retention. What will you do to keep customers coming back to you for service? Tools such as customer surveys and promotions will keep customers engaged in your business operations.
- **Community:** Producing income for your business is certainly a main goal for your business, but giving back to the community through sponsorship and volunteer work is another important goal. Decide early which community groups or organizations you would like to support.
- **Revenue:** Set a revenue goal for your business. Determine how much you would like to make in service and retail sales, as well as how much you would like to earn in profit each year (beyond what you pay yourself).
- **Expansion:** Create goals related to the growth of your business. Determine what services or products you would like to add as your business grows. In this area, determine if your goal is to open a full-service salon. If this is your plan, make sure you detail what it would take to accomplish this goal.

FIGURE 3–5

Write Down Your Long-Term Goals.

It is imperative to keep a detailed journal of your short- and long-term goals so you can visit them often. This is important, because when you actually start performing the day-to-day activities of your business, you often experience new and exciting things that can change your short- and long-term goals. When you encounter these new experiences that may trigger a change in your goals, you want to make sure you write them down so you can stay on the right path, because it is easy to get sidetracked. Use Exercise 3–2 to write down your short- and long-term goals.

RESOURCES

Web-based software programs like BizOn Track are great goal-management tools to help you manage your goals and keep you on track. Visit http://bizontrack.com for more information.

EXERCISE 3–2 Write Out Your Short- and Long-term Goals

SHORT-TERM GOALS	LONG-TERM GOALS	DATE TO BE COMPLETED
1.	1.	
2.	2.	
3.	3.	
4.	4.	
5.	5.	
6.	6.	
7.	7.	

3.3 Identify Your Core Business Values

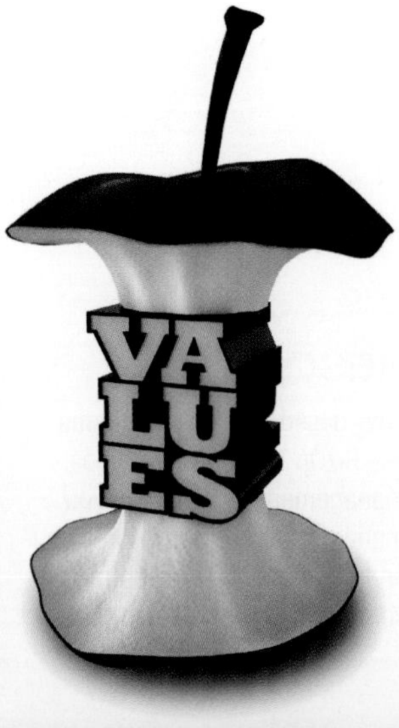

Core business values are defined as a set of principles and beliefs that your company is founded upon and that represent what your company believes (Figure 3–6). As a booth renter, defining your core values is important because, since you are independent, it will mirror your personal core values. When creating your business core values, you want to narrow them down to about three to five important core values that are important to you.

Let us do an exercise that will help us create our core value statements. Select the top five value principles from the list in Exercise 3–3 that matter to you the most.

Now that you have chosen some principles that matter the most to you, it is time to develop your core value statements. A core value statements describes how you intend to "live out" that value in your business. Let us explore how to create your core value statements.

FIGURE 3–6
What Are Your Core Values?

EXERCISE 3–3 Core Values List

Circle the top five value principles that matter to you the most. Feel free to add or exchange the qualities or characteristics listed in the following.

Family	Success	Power	Fame
Wealth	Peace	Happiness	Joy
Truth	Wisdom	Justice	Love
Integrity	Honesty	Loyalty	Respect
Quality	Service	Freedom	Courage
Authenticity	Dependability	Excellence	Spirituality
Learning	Relationship	Trust	Persistence
Accountability	Security	Strength	Purpose

Your first step is to answer the following questions for each of the five core values you have selected. Here is an example using the principle "excellence":

- *What does the principle mean to you?* Excellence is defined as excelling, offering good quality, and being the best at what you do. It is also defined as working at your optimal level.
- *What do you believe?* Every customer deserves excellent customer service.
- *What do you think?* Every person that receives a service from me should leave satisfied.

- *What do you do?* We keep ourselves up to date with the latest in hairstyling techniques and tools by participating in ongoing education.
- *What is the evidence?* Our customer retention rate continues to rise with customers rebooking for future appointments at the end of their service.

Based on the information provided, your first core value statements may be written as follows:

XYZ Salon Service is dedicated to supplying excellent customer service by providing clients with the latest in hairstyling and salon technology.

Now it is your turn! You will want to answer the same questions as in the example to help you create your own value statements.

- What does this value *principle* mean to you?
- What do you believe?
- What do you think?
- What do you do?
- What is the evidence?

Use Exercise 3–4 to create your top five core value statements. Use the five core value principles you selected in Exercise 3–3 as a guideline for creating your core value statements.

Establishing your core values and writing statements for each is an important step toward understanding exactly the type of business you intend to build. Now that you have done the work on values, it

EXERCISE 3–4   Write Your Company's Top Five Core Value Statements

Value Principle	Core Value Statement
	1.
	2.
	3.
	4.
	5.

is time to write your vision and mission statements in Exercises 3-5 and 3-6. With your core values fresh in your mind, begin to draft each statement. You may even choose to use some of your values in the statements themselves. Get creative and have fun with it. You will know when you have it right, because when you read each one, you will feel a sense of pride and excitement.

EXERCISE 3–5 Write a Mission Statement for Your Business

_____
_____
_____
_____
_____
_____

EXERCISE 3–6 Write a Vision Statement for Your Business

_____
_____
_____
_____
_____
_____

3.4 Create a Solid Business Plan

Earlier in the chapter, we discussed the difference between a business plan and a business model. A business plan gives you specifics about your business, while a business model defines how your business will make money. Although you are running an independent booth-rental business, it is still both important and necessary to create a solid business plan to compliment your business model. Doing so will help you clearly see what it would take to successfully operate your booth-rental business.

The business plan will spell out the specifics of where you plan to service your customers, how you plan to market for new customers, and the financials of your business (Figure 3–7). When creating a business plan for your business, SBA.gov recommends your business plans include the following elements:[4]

A. **Business Plan Executive Summary:** This is a descriptive summary of your company that includes your mission statement, company information such as the date your business was formed, who the owner(s) are and their titles, the business location, the number of employees, the description of your services, financial growth numbers (if available), and where you plan to take your company in the future.

B. **Market Analysis:** In this section, you will provide a brief summary about the cosmetology industry.
- Provide industry data about salon employees versus booth renters.
- Describe who your customers are.
- Define your pricing structure. Make a list of your services and define what the price of each service will be.
- Provide a competitive analysis. A **competitive analysis** points out your competitors and takes a look at their business tactics to identify their strengths and weaknesses compared to your business.[5] Providing a competitive analysis is great, because it will help you determine your pricing structure based on other businesses in the area.

C. **Company Description:** Describes your business and how the services and products you offer will meet the needs of your target market. It will also describe what is different about your company and list your competitive advantages.

> **COACHING NOTE**
> Complete your executive summary last. Once you have clearly identified all of the minor and major details of your business, you can then "sum" them up in your executive summary. An executive summary is the first thing that is read by a lender or investor. Be sure it is concise and that it thoroughly captures how you will succeed.

FIGURE 3–7

It Is Important to Create a Solid Business Plan.

D. **Organization and Management:** This section will talk about how your company is structured. It will include

- An organizational chart (optional if you are the sole employee).
- A description of the corporate structure, and whether your business is a sole proprietorship, a limited liability company, an S corporation, or a C corporation.
- The name of the owners, the percentage of the company owned, and a biography of the owners that includes their position, work experience, education, prior employment, special skills, industry awards and recognitions, and compensation.

E. **Marketing Plan:** This section will describe how you plan to gain new customers. It will describe specific market strategies and tactics that will be used to grow your business.

F. **Service and Products:** This section will describe the services you will offer and how these services will benefit your customers. Also include any legal contracts, agreements, copyrights, patents, or trade secrets in this section.

G. **Funding Request:** If you are looking for financial assistance, this section will provide information such as the amount of money you are asking for, what the money will be used for (e.g., working capital, debt pay off, etc.), and the length of time desired to pay the loan off.

COACHING NOTE
If you are seeking financial assistance from a lender to open your booth-rental business, you must show "how" you will make money, as well as outline a history of income for him or her. This is another reason why accurate tracking, recording, and reporting of your earnings are essential.

H. **Financial Projections:** This section will provide data about your projected income, detail what you expect to make, how you expect to make it, and what the expenses are to determine your gross profit. We will discuss finances more thoroughly in Chapter 4.

Creating a solid business plan is a crucial part of creating the type of business you deserve. It is recommended you contact your local small business administration or an industry coach to help you successfully design your business plan.

Fill out the business plan template in Exercise 3–7 to get you started with the information you need to create your business plan. Do not worry if you cannot complete it all right now. Continue reading the entire book and you will find exercises and answers as to what you may want to include in your plan.

EXERCISE 3-7 — Booth Renter Business Plan Template

Business Name:

Hours of Operation:

Location:

City Code: State: Zip code:

Executive Summary:

Market Analysis:
  Industry Data:

Describe Your Customers (Demographic):

Price Structure:

(In this section, list your services and the price you will charge for the service.)

Service Name	Service Price
Shampoo	$
Cut	$
	$
	$
	$
	$
	$
	$
	$
	$
	$
	$
	$
	$
	$

(Continued)

EXERCISE 3-7 Booth-Renter Business Plan Template (*Continued*)

Competitive Analysis Chart:

Use the competitive analysis chart to compare your business to your competitors. Rate each factor as good, fair, or poor.

FACTORS	YOUR COMPANY	COMPETITOR 1	COMPETITOR 2
Location			
Environment			
Services Offered			
Prices or Services			
Target Customer			
Amenities			

Company Description:

Organization and Management:

Corporate Structure:

Owner: Title:

Percent of Company Owned:

Education:

Number of Years in the Business:

Special Skills:

Awards and Recognitions:

Compensation:

(Define what your compensation will be—for example, 100 percent of earned services or flat-rate pay.)

Owner's Biography *(use additional paper if needed)*:

Marketing Plan

Marketing Strategy 1

Marketing Strategy 2

Marketing Strategy 3

Marketing Strategy 4

Marketing Strategy 5

Funding Request

Will you need to borrow money to start your booth-rental business?

If yes, how much money do you need? $

What will you use the money for?

How long are you requesting for payback?

Financial Projections

What is your projected monthly income:? $

Describe how you plan to generate this income?

EXERCISE 3-7 — Booth-Renter Business Plan Template (Continued)

Basic Business Monthly Expenses:

Item	Cost
Booth-Rental Fees	$
Product Purchases	$
Marketing/Advertising Costs	$
Bookkeeping Fees	$
Credit-Card Terminal Fees	$
Phone Service Fees	$
Internet Fees	$
Website Fees	$
Assistant Pay	$
Assistant Pay	$
Receptionist Pay	$
Towel Service Fees	$
Cleaning Fees	$
Professional Liability Insurance	$
Disability Income Insurance	$
Continuing Education	$
Misc. Business Expense	$
Misc. Business Expense	$
Misc. Business Expense	$
TOTAL EXPENSES:	$

3.5 Growing the Business

The one thing you have to keep focus of for any business plan is growing your business. Yes, you are booth renting, but it is still a business that has room for growth. As you create your short- and long-term goals, it is important to also create an annual income goal for your booth-rental business. Setting an annual income goal will keep you focused on the goal of growth for your business, which in turn will keep you focused on keeping your pricing structure for services stable. We will talk more about setting income goals in Chapter 4, "Your Money, Your Future."

If you want to grow your booth-rental business, follow these 10 steps:

1. **Charge what you are worth:** In the service industry of cosmetology, the way a stylist makes his or her money is by performing services on customers. One of the biggest problems some independent stylists have is in pricing their services correctly. When you have an improper pricing model, it causes you to struggle financially or do things like overbook clients just to make a decent wage. Determining pricing involves a great deal of thought and assessment. Once you have considered all of the factors listed next, you can confidently establish a pricing menu that is fair, justified, and profitable for the business you desire. Things to consider when creating a price structure for your services are as follows:

    a. **Location:** One thing to consider is your location. Does your location have easy access with adequate parking?

    b. **Salon Amenities:** Compare the amenities offered at the salon you are renting from, or plan to rent from, with other salons in the surrounding area that offer the same services. For example, does the salon you rent from offer electric shampoo chairs, an espresso/latte machine, and free WiFi? If so, this should be taken into consideration when pricing your services.

    c. **Supply and Demand:** Truth be told, no matter how good you are, if you are not booked 70 to 80 percent of the time, you have not earned the right to charge higher rates, regardless of your skill, the salon amenities, and your desired income. Setting your prices too high before you have created a demand will only leave you sitting around "hoping" people show up. If there is a high demand for your services, and you find yourself booked four to six weeks out, then it may be time for you to consider a price increase.

    d. **Experience and Expertise:** Consider your experience and expertise in the beauty industry. Experience does play a part in pricing your services, but do not mistake experience for tenure in the industry. There are plenty of professionals out there who have taken very few advanced education courses throughout their career. So, although they may have 15 years in the industry, they do not necessarily have 15 years of experience. Thoughtfully consider whether or not you have completed enough advanced education and additional certifications. Consider the following things:

        - Do you have experience as an educator with a product-manufacturing company?

- Are you an independent beauty professional who has worked in film or television?
- Have you worked on celebrity clientele?
- Do you consider yourself an expert in certain services?
- Are you a new stylist with little to no experience?

Whatever the answers are to these questions, use the information from each point to help you determine an appropriate pricing menu.

e. **Personal Income Goals and Expenses:** Finally, the most important thing to consider when establishing your pricing is your desired income as well as your personal expenses. Think of it this way. Let us say that you set your pricing after considering all of the preceding factors—experience, expertise, productivity, amenities, and location. You even considered what the other renters in your salon or in salons around you were charging. Now, imagine a month has gone by. Reflecting back, you were happy during the month. Your clients seemed to enjoy your new environment, you were proud of what you were doing, and your productivity was well over 75 percent. However, your revenue earned for the month was only $4,000 and your monthly personal and professional expenses exceeded $4,000. You basically spent all that you earned just to cover all your expenses. That, I am sure, would not make you happy. What happened you ask? Unless you were giving away your services for free, the only conclusion you can come to is that you did not establish your pricing based on your personal and professional expenses. Use Exercise 3–8 to help you determine a starting point for your pricing.

EXERCISE 3–8 Base Price Worksheet

This worksheet is designed to help you calculate your "base price." The "base price" is the price you charge for the most commonly purchased service in your business. It is the starting point for all other service pricing. If you are a hairdresser, and your haircut is your most commonly purchased service, then the price established is what you charge for your basic haircut.

The "base price" calculation tells you *what you minimally must charge each client in order to cover your expenses and reach your desired income.*

Step 1) ENTERING THE INFORMATION:

Line 1) TOTAL <u>MONTHLY EXPENSES</u> $_____
(enter your TOTAL monthly personal and business expenses on this line)

Line 2) PROJECTED <u>MONTHLY PROFIT</u> $_____
(enter the amount of money you WANT to make each month on this line)

Line 3) <u># OF MONTHLY CLIENTS</u> $_____
(enter your average client count for a month on this line)

Step 2) CALCULATING THE INFORMATION:

Line 4) TOTAL <u>PROJECTED GROSS SALES</u> $_____
(take the "Total Monthly Expenses" (line 1) and ADD to the "Projected Monthly Profit" (line 2) and put the answer on this line)

Line 5) <u>RECOMMENDED BASE PRICE</u> $_____
(take "Total Projected Gross Sales" (line 4) and divide by the "# of Monthly Clients" (line 3) and put the answer here)

Keep in mind that it is a smart business practice to determine the price point that ultimately will ("on paper") help you reach your desired income. The operative words here are "on paper."

Remember, as we discussed earlier, this is one component of determining your pricing.

The results of your worksheet reveal several truths:
1. *My prices are fine. If this truth is revealed, ask yourself if FINE is where you ultimately want to be? If not, shift your goals and take the necessary actions.*
2. *My prices are too low.*
3. *My prices are too high.*

There are several remedies to these truths.

<u>ONE:</u> **Raise Prices**
If you have established a demand for your business, are busy, and/or have a waiting list, then what are you waiting for? Raise your prices! Your fear of losing clients is killing your opportunity for profitability. Again, ask yourself, am I doing the necessary actions to achieve my dreams? Meaning… am I providing the consistent level of service that supports my pricing. If the answer is no, then act!

<u>TWO:</u> **Up Service**
Increasing your average ticket may seem simple at first, but in reality it often requires more work and effort than an individual is willing to apply. The good news is that you now have established the minimum each client is to spend at each visit. The bad news is that you have got to somehow get them to spend that amount consistently—if they do not, you still will not reach your minimum base price and continue to remain unprofitable.

<u>THREE:</u> **Reduce Expenses**
Although obvious, reducing expenses is definitely easier said than done. Doing so requires discipline, accurate recordkeeping, tracking, and analysis. QuickBooks or a similar expense tracker will support you in your recordkeeping. Next, you may need to consider hiring or replacing an accountant. Finally, it may be time to reign in your credit–card debt. Contact a debt consolidation company to help you organize and reduce your debt.

<u>FOUR:</u> **Increase Client Count**
Making an effort to increase your client count may be necessary. Establishing a clear and written three- to six-month marketing plan (see Chapter 5) to increase the number of new clients is an important next step. Be sure to track your results so you can evaluate and make the necessary adjustments before engaging in additional efforts to increase your client count.

(Continued)

> **EXERCISE 3–8** Base Price Worksheet (*Continued*)
>
> If your prices are too *high*, it can be difficult to reach profitability because you have not established a solid market demand, and so you will struggle to obtain enough customers to generate the necessary revenue. Wake up—your ego may kill your opportunity for success. If you are not booked over *80 percent* of the time, then you have not earned the right, or established the necessary demand, to charge the prices you have set.
>
> Regardless of your predicament, both scenarios can be solved by understanding basic business economics: supply versus demand, and income versus expense. When there is more supply than demand, lower your prices and build your brand to increase your market share. If demand is greater than the supply, then raise your prices and earn what the market is telling you that you are worth. Conversely, obtaining an accurate analysis of income versus expenses will ultimately provide the framework necessary for establishing price points. Setting your prices *right* the first time will eliminate guesswork and unnecessary and often long-term revenue losses.

Armed with the knowledge from the Base Price Worksheet, you can confidently and as accurately as possible establish your price menu. Plan to make this self-reflection, analysis, and price calculating an annual event. Doing so will ensure that you can raise your prices, and give yourself a raise, as often as needed.

Income growth can be challenging to booth renters because oftentimes renters become so engrossed in the styling side of the business that they forget to run the financial side of it. Thus, the stylists operate for several years without increasing the prices of their services. The lack of service price increases will have some independent stylists working but still suffering financially. There have been cases where stylists did not increase their prices for 10 years or more. Most often, the reason for this lack of price increase is because stylists fear they will lose their clients if they raise their prices. In order to have income growth, you must give yourself a raise by increasing your service prices. By following the earlier exercises, you can confidently do this when your business "says" so, not when you feel like it (Figure 3–8).

Another thing to keep in mind is that the prices of the products you use to perform these services increase yearly, as well as the tools you use to perform hair services. So why haven't you increased your prices? It seems only logical.

FIGURE 3–8

Charge What You Are Worth for the Services You Provide.

The clients you service work their jobs in expectation of receiving a cost-of-living increase every year, so why not you? It is common practice for companies to give their employees an annual cost-of-living increase of about 3 to 5 percent. As a self-employed, independent beauty professional, this is your job, and you deserve to give yourself a raise. The only difference between you and your clients is that you have to give yourself a raise by increasing your prices.

Whenever you raise your prices, be sure to inform your clients in advance. Often times, it is not the price increase that causes clients to leave. It is usually a result of poor service, inconsistent pricing, and a failure to inform them of changes. You can opt to send a letter or simply let each client know at the end of the service that upon their next visit there will be a slight price increase. Be confident and professional; do not "explain" your increase. Simply state what the new price will be and move on. The less of a deal you make of it, the less your client will as well. If you have done the work and know that the signs all point to a price increase, then there is no need to stress about it. You may lose one or two clients—that is expected—but you will make more on those who remain.

2. **Accept all forms of payments:** One way to grow your business is to accept credit/debit cards from your customers (Figure 3–9). Accepting credit and debit cards from your customers as an independent beauty professional can increase your revenues greatly. When customers are not restricted to a cash amount for spending, they will often add additional services or retail items to their ticket that they may not have originally planned on—but all of which will increase your total service ticket considerably.

3. **Create an atmosphere for spending:** Have you ever noticed that when you go to the mall, an atmosphere has been created there that lets you spend money? The retail displays are appropriately signed, mannequins are dressed to get your attention, music plays softly, and the décor draws you into a mode of spending. These are the same components you should create in your salon environment. Choose a salon booth to rent from whose

FIGURE 3–9

Accept Several Forms of Payment.

FIGURE 3–10
Select a Work Environment That Is Conducive For Spending.

atmosphere will make a client to want to spend money (Figure 3–10). Also, set a small retail display at your station that is attractive, or use table top acrylic signage with a nice graphic that promotes a service or product. If these components are used properly, your customers will spend!

4. **Retail:** Another key factor in helping you grow your booth-rental business is to retail. Retailing can make up 15 to 30 percent of your income. Since retailing is very important for the growth of your business, it is covered more deeply in Section 7.3 of Chapter 7, "The Day-to-Day Details."

5. **Be professional:** In order to grow your booth-rental business, you must look like you are ready to provide excellent service to your clients. When you dress for success, you will attract the type of clients who will pay you what you are worth (Figure 3–11). Dress professionally! Do not wear clothes that look like you went out partying the night before. Do not start a service on yourself and then greet your clients with your hair untamed and having rollers or haircolor in your hair. If you want to increase your income, dressing for success is a definite must.

6. **Keep up with the latest hair trends:** Keeping up with the latest hair trends will definitely increase your income. The most popular service with the largest hair ticket is hair extensions. If you provide hair extension services, you will definitely see an increase in your profits, and you will be well on your way to seeing an income growth in your business.

7. **Know your stuff:** Make sure you keep abreast of the latest news in the beauty business. Attend classes and trade shows to keep up with the latest in style trends and new hair technology. Also, stick to what you know! If you are not good with updos, do not try to do them. Stick to the services you are best at. If you try to perform services that you are not an expert in, you can actually damage your business as an independent beauty professional. If you perform a service that does not have an attractive finish, clients will leave saying that

HERE'S A TIP

Many salons that offer booths for rent will tell the stylists that they are not allowed to sell retail as a booth renter because they have a retail product in the salon that they want to sell. Be very aware that this type of statement or activity borders on treating you as an employee, which is defined as someone who is controlling your work activity. So make sure you do not sign any contracts or agreements with this verbiage, unless you wish to sell the retail item offered in the salon.

you are not a good stylist, when this may not be the truth. Perhaps you are not great in styling updos, but you are excellent in cutting and coloring, so stick to what you know.

8. **Keep a positive attitude:** Keeping a pleasant attitude is another key attribute to growing your business. Having an attitude of gratitude and appreciation toward your clients will increase your clientele. One way to show appreciation to your customers is to give appreciation gifts during the holiday season. For example, if you love to talk about movies with your clients, during the holiday season give your top 25 paying customers a $5 movie gift card. Customers like this type of appreciation, and they will refer other customers to your website and to you, just because you have the right attitude.

FIGURE 3–11

Come to Work Dressed for Success.

9. **Create a successful marketing plan:** Creating a successful marketing plan is also a key to growing your business. We will cover more on creating a successful marketing plan in Chapter 5, "Marketing Your Booth-Rental Business."

10. **Get an assistant:** Lastly, one key to growing your booth-rental business is to hire an assistant. It is hard to grow your business if you do not have an assistant. Many booth renters have questions about how much money you should pay an assistant. You can pay your assistant anywhere from $8 to $10 an hour, depending on the minimum wage per state. Hiring an assistant to help you with the servicing of your customers is a big plus, but before you hire an assistant, make sure you find out the rules and regulations in your state about what services an assistant can perform (Figure 3–12). Also, you will need to find out if your state requires that an assistant have a shampoo assistant license.

> **HERE'S A TIP**
>
> If you are not great at performing a certain service, partner with another stylist whom you can refer your clients to. Your clients will appreciate the referral and the great customer service you provide.

FIGURE 3–12

Hire an Assistant If Necessary.

3.6 S.W.O.T. Analysis

As you work to grow your independent beauty business, keep in mind that your growth and success is the number one goal. In order to grow and be successful, it is important to know your company and its industry. In the business world, the way to identify the characteristics of both of these is to perform a S.W.O.T. analysis on your business. A **S.W.O.T. analysis** is a planning method used to identify the *strengths*, *weaknesses*, *opportunities*, and *threats* of your business[6] (Figure 3–13). A S.W.O.T. analysis will help you to recognize those internal and external factors that can help you or delay you in achieving your business goals. Generally when you identify the strengths and weaknesses of your company, these are internal characteristics. Identifying your opportunities and threats has to do with factors located outside your company.

Before you perform a S.W.O.T. analysis on your business, answer the following questions categorically.

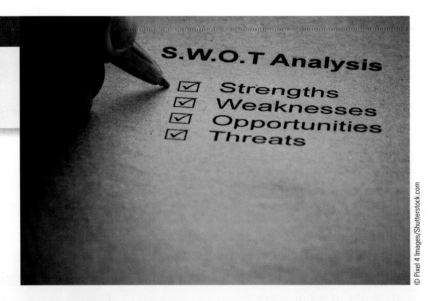

FIGURE 3–13

Use a S.W.O.T. Analysis to Help Determine Your Company's Strengths and Weaknesses.

Strengths

When thinking about your strengths, look at them from three perspectives: the internal perspective, customer perspective, and peer perspective. Strengths can also include providing cutting-edge technology when servicing clients, as using the latest products, services, and hair tool technology.

- What does your business provide that is an advantage over other businesses providing the same service? _____
_____
_____

- What service do you do best? _____
  _____
  _____

- What sets you apart from others in the industry? _____
  _____
  _____

- What do your peers in the industry see as your strengths? _____
  _____
  _____

- What makes your customer want to pass other salons and come to you to receive a service? _____
  _____
  _____

Weaknesses

When thinking about your weaknesses, also look at them from three perspectives: the internal perspective, customer perspective, and peer perspective. Weaknesses can also include the use of technology and latest trends.

- What things can you do to improve your services? _____
  _____
  _____

- In what area of your business do you need the most help? _____
  _____
  _____

- What feedback have you received from your customers that identify room for improvement? _____
  _____
  _____

- What things have happened that made you lose revenue? _____
  _____
  _____

Opportunities

Look at your strengths to determine if you can create new opportunities, and look at your weaknesses to determine if you remove a weakness, will any new opportunities await you.

- In your field of service, are there any opportunities that are wide open in your area? _____
_____

- What new products, services, or hair technology are you aware of that are not being offered in your area? _____
_____

- In what market demographic do you see the most opportunities?
_____
_____

Threats

- What challenges does your business face? _____
_____

- What are your competitors doing that you are not doing?
_____
_____

- Are there changes in the industry or domestic culture that are having an effect on the way you do business? (For example: the struggling economy) _____
_____

- Have you incurred debt that you are having problems paying?
_____
_____

- Are you having cash flow issues? _____
_____

- Is your pricing structure too low or too high, and is it having an effect on you building clientele? _____
_____

- How is your salon environment compared to other salon environments in the surrounding areas? _____
_____
_____

Use the S.W.O.T. analysis chart in Exercise 3–9 to perform a S.W.O.T. analysis on your company.

EXERCISE 3–9 S.W.O.T. Analysis

STRENGTHS	WEAKNESSES
○	○
○	○
○	○
○	○
○	○
○	○
○	○
○	○

OPPORTUNITIES	THREATS
○	○
○	○
○	○
○	○
○	○
○	○
○	○
○	○

3.7 Summary

Creating a solid business model and plan for your business before you start your operations can help you avoid a lot of costly mistakes and challenges that many booth renters face. If you create a firm business model in the beginning, you will create a path for business growth and success. Make sure your business plan is detailed with operation specifics to ensure you have all the facets of your business model covered.

Let us summarize the top valuable lessons you have learned from this chapter to help you eliminate detours and distractions while building your business.

3.8 Top Takeaways: Creating a Solid Business Model

- **Business model.** Create a business model for your business, because it will define a clear roadmap as to how your business will make profits.

- **Mission statement.** A mission statement is important to create because it will detail your company's purpose and reason for existence.

- **Vision statement.** Creating a vision statement will describe your company's futuristic goals, values, and direction.

- **Short-term goals.** Setting short-term goals for your business will give you some goals to achieve within the first two years of your business.

- **Long-term goals.** Setting long-term goals will give you business objectives to achieve within three to five years.

- **Pricing structure.** Plan to grow your business successfully by using good pricing systems to create your service pricing menu.

- **S.W.O.T.** Perform a S.W.O.T. analysis to determine your business's strengths, weaknesses, opportunities, and threats.

---

[1] The U.S. Small Business Administration. (n.d.). *What is a business plan and why do I need one?* Retrieved from http://www.sba.gov.

[2] Business Model Definition Investopedia. (n.d.). *Investopedia – Educating the world about finance.* Retrieved from http://www.investopedia.com.

[3] Media, D. (n.d.). *What are the main components of a business model?* Retrieved on April 8, 2013, from http://smallbusiness.chron.com.

[4] The U.S. Small Business Administration. (2012). *How to write a business plan.* http://www.sba.gov.

[5] Entrepreneur Media, Inc. (2012). *Definition of competitive analysis.* Retrieved from http://www.entrepreneur.com.

[6] Mind Tools - Management Training, Leadership Training and Career Training. (n.d.). *SWOT Analysis - Strategy Tools from MindTools.com.* Retrieved from http://www.mindtools.com.

Chapter 3 Quiz: Eliminating Detours and Distractions with a Solid Business Model

This chapter described the importance of having a business plan when starting your booth-rental business. Answer the following questions to review what you have learned.

1. The purpose of a mission statement is to _____.
    a. talk about your business structure
    b. discuss your pricing structure
    c. explain why you decided to open your business
    d. all of the above

2. A vision statement _____.
    a. is a short-term picture of your business
    b. states the concerns you have for your company in the years to come
    c. specifies where you want your company to be in the years to come
    d. all of the above

3. List two types of goals you should have for your business.
    1. _____
    2. _____

4. Your short-term goals should be S.M.A.R.T. Spell out the acronym for S.M.A.R.T.
    S _____
    M _____
    A _____
    R _____
    T _____

5. Your long-term goals should include _____.
    a. service
    b. community
    c. revenue
    d. expansion
    e. all of the above

6. A business plan and a business model are the same.
    a. True
    b. False

7. A business plan defines how your business will make money.
    a. True
    b. False

8. A _____ gives you specifics about your business.
   a. business plan
   b. business model

9. When creating your pricing structure, you should consider _____.
   a. location
   b. salon amenities
   c. experience
   d. all of the above

10. List five things that will help your business grow.
    1. _____
    2. _____
    3. _____
    4. _____
    5. _____

11. List four components of a S.W.O.T. analysis.
    1. _____
    2. _____
    3. _____
    4. _____

chapter 4

Your Money, Your Future

CHAPTER OUTLINE

4.1 Costs to Succeed

4.2 Evaluate Your Personal Expenses

4.3 Establish Income Goals

4.4 Establish Your Financial Dream Team

4.5 Create a Financial-Health Checklist

4.6 Analyze Your Financial Health

4.7 Health Benefits

4.8 Taxes

4.9 Saving for Retirement

4.10 Build Business Credit

4.11 Summary

4.12 Top Takeaways: Your Money, Your Future

Career Profile

Ramin Romney

Ramin Romney has been in the industry for 11 years, and was a booth renter for 2 years before becoming a business owner, so he can offer advice from the perspective of both an owner and booth renter. In 2000, after gaining experience working for another barber shop, he opened his own shop "Set N Trends" in Jacksonville, Florida. He offers a booth-rental structure on a weekly basis. In addition to working behind the chair and being a shop owner, he has served on various boards, such as the Fortis Institute Advisory Board, and has been featured in professional showcases at the Bronner Brothers Hair Show in Atlanta, Georgia, with Craig Damon, who is actor Tyler Perry's personal barber.

What is the structure of your booth-rental business?

The barbers who rent booths are considered independent contractors because they are responsible for maintaining their own clients, revenues, expenses, income taxes, insurances, and the like. The booth-rental arrangement allows shop owners to minimize expenditures on their own overhead while providing the renter with the opportunity to experience the real costs and considerations of business ownership on an individual level. As a shop owner, I prefer the booth-rental arrangement because of the fixed rate—in my area, they average between $100 to $160 on a weekly basis, depending on the shop. Having a fixed rate simplifies the monthly budgeting and forecasting of the shop's income and expenditures.

How do you keep all of your finances in order? Any advice for those who cannot afford to pay a full team of help, such as a bookkeeper and so on? Do you use an accountant?

My finances are kept in order by using my business checking account or credit card for all transactions pertaining to the business; therefore, I am able to account for all expenses paid out for the business. If for some reason an item has to be paid with cash, I keep all the receipts so I can write off those items when doing taxes. I have an accountant to help keep track of taxes, but the best way to save money even with an accountant is to keep track of expenditures yourself; therefore, the added expense of a bookkeeper is not needed. An app that I currently use for my phone is a credit/debit swiper that processes payments at a lower fee than conventional credit/debit machines. There is no special software needed to track finances as long as your transactions are linked specifically to your business bank account.

How do you handle scheduling work hours? Do you create a flexible schedule based on your client's needs?

Each hair cut is scheduled in 30- to 35-minute intervals so there are no overlaps in customers. My schedule is flexible based on my customers' needs, because most people work during the day. So you have to be available if they want to come in early, on their lunch break, or when they get off of work.

What do you want to share with future booth renters about evaluating finances?

My advice for booth renters is to keep a receipt book with the weekly booth rent to show the money spent on your space, something which can be used later in listing your tax expenses. For personal finances, it is important to pay yourself and set income goals per month/year based upon your needs. Fortunately, at this time there are many individual policies available for health, dental, and vision plans, so the days of no health care for independent contractors are no longer, due to the many affordable policies available. A retirement plan will differ per the individual, but some options that should be looked into would be life insurance policies and 401K/IRA/Roth plans. Business credit can be established in multiple ways, such as maintaining business expenses, loans, and credit cards.

Running a booth-rental operation can be financially challenging, but it can also be financially rewarding. The greatest financial challenge booth renters face is combining their personal finances with their business finances. In most cases, renters will integrate business income and expenses with personal income and expenses, although it is very crucial that the two incomes stay separate. In this chapter, we will discuss what it takes to financially run your booth-rental business.

4.1 Costs to Succeed

When the question is asked about how much money it would take for you to succeed as an independent beauty professional, this answer is directly related to what your personal expenses and business expenses are collectively. If you recall Exercise 1–1 in Chapter 1, we presented a booth-renter readiness chart that gave you a quick summary of the business expenses you would have as a booth renter. In Exercise 1–2, we presented an income-and-expense worksheet that gave you a quick summary of what your personal expenses are. If you combine the total expenses from Exercise 1–1 with your total expenses in Exercise 1–2, this will give you a quick synopsis of the revenue you need to generate.

The good thing about running a booth-rental operation is that it really does not require a lot of overhead expenses. The basic expenses required to run the business are mainly your booth rent and the supplies needed to service your customers. But your business's financial success will depend on these five items: cash flow, building clientele, retaining clients, scheduling/available work hours, and personal expenses.

Cash Flow

Cash flow is defined as the amount of money coming in as revenue and going out of your business for expenses (Figure 4–1). If you have more revenue coming into your business than money going out for expenses, this is called **positive cash flow**. If you have more money going out of your business to pay expenses than you have coming in, this is called **negative cash flow**. This is where most booth renters get in trouble. Although running an independent booth-rental business does not require a lot of overhead, you will fall into financial woes if you have more money going out of the business than coming in. Since the income that a booth renter generates has to cover business expenses and personal expenses, it is important

FIGURE 4–1

Your Business Success Depends on the Cash Flow.

> **HERE'S A TIP**
>
> **Cash Flow**
> One thing you can do to help you with your cash flow is to use a credit card to pay for expenses like product supplies, retail inventory, educational classes, and marketing fees. Obtain a separate credit card for your business, in your business name, that you will use only for business expenses. This will help increase your cash flow as well as help you keep track of your business expenses.

when creating your income goals to consider both your business and your personal expenses in your income-goal projection.

Scheduling Work Hours

Flexibility in the hours you are able to work will impact cash flow and is another key to success (Figure 4–2). It is hard to build a list of clientele when you are not available to work during peak service hours. This is a major factor in the success of generating revenue for your business. You must have working hours available that meet the needs of the customers. If you have customers who work in the daytime, design a work schedule that will accommodate servicing these clients after work or on the weekend.

When creating your work schedule, keep in mind the needs of your clients. This is important to mention because some booth renters' hours of operations are designed to fit their own personal needs instead of the needs of the clients. Let us look at an example in the Coaching Note below of a work schedule that is self-serving.

The moral of the story is that when you are building clientele, you need to have flexible hours when you are available to service the needs of the clients. If you are not available when the clients need you, the clients will find someone else who *is* available. This can impact your cash flow, interrupt your ability to build client partnerships, and limit the revenue for your business.

Personal Expenses

Personal expenses play a major part in how much you will need to succeed. If your personal expenses are high, then your prices may

> **COACHING NOTE**
>
> Jill is a booth renter and is married with three children. Her children are elementary school–aged, and Jill is responsible for getting them to school in the morning and picking them up in the afternoon. Jill's sister picks the children up from school on Mondays, Tuesdays, and Wednesdays, but it is Jill's responsibility to pick them up on Thursdays and Fridays. As an independent booth renter, Jill is responsible for setting her own hours of operation. Since she has personal responsibilities, she sets her hours for servicing clients on Mondays, Tuesdays, and Wednesdays from 9 am to 5 pm, and Thursdays and Fridays from 9 am to 1 pm. She is off on Saturdays. The challenge with this work schedule is that she is not available to service clients during the beauty industry's peak business hours, which are in the evenings, during the later part of the week and on weekends. If Jill's schedule does not allow for the flexibility of working during peak hours, Jill will miss the opportunity to service a whole group of customers who visit the salon frequently during these peak times. This schedule will hurt Jill's revenue production, especially if Jill is trying to build clientele.
>
> Now, if Jill is a seasoned stylist with years of experience and a strong clientele with whom she has made strong partnerships, these hours will work because she would have already built strong client partnerships and a demand for her services that will allow her schedule to be filled during the hours of operation she has set.

need to increase and you will need to set a higher income goal than you would if your expenses were lower.

As you can see, much of what you need to succeed as a booth renter financially has to do with your personal expenses.

4.2 Evaluate Your Personal Expenses

Since we have discovered that much of what you would need as a booth renter to succeed is tied to your personal expenses, let us evaluate your personal expenses (Figure 4–2).

The best way to evaluate your personal expenses is to create a household budget. Creating a household budget allows you to clearly define how much money you are bringing in and how much money is going out. In Chapter 1, Exercise 1–2, you summarized your total monthly expenses; now let us find out if your personal cash flow is positive or negative. Complete Exercise 4–1 on the following page to evaluate your personal cash flow.

Now that we have a summary of your total cash flow, we can establish some income goals that you will need in order to succeed.

FIGURE 4–2

Create Available Hours That Fit Your Client's Needs.

4.3 Establish Income Goals

Establishing an income goal is a great way to help you stay on track with your revenue-generating efforts (Figure 4–3). When setting your income goals, you should look at how much revenue you need to generate to cover all of your personal and business expenses. The task you completed in Exercise 4–1 should have given you a clear idea of how much money you need to generate monthly. Since most booth renters pay their booth rent weekly, we are going to create weekly income goals and daily income goals from your annual income goal. Creating a daily income goal based on your monthly income goal will give you a goal to set daily to keep you focused on the annual goal.

In order to create your weekly and daily income goals, you will need to record some information. Please answer the following questions so we can begin creating our daily income goal.

- What is your desired annual income goal? $ _____
- How many days do you work in a week? _____

FIGURE 4–3

Set Up Your Income Goals.

EXERCISE 4–1 Personal Cash-Flow Evaluator

Fill out the information in Exercise 4–1 to evaluate your personal cash flow.

A	Total Monthly Income	$
B	Income from Other Sources (including tips)	$
C	**TOTAL INCOME** (add rows A + B)	$
D	Total Monthly Personal Expenses from Exercise 1–2	$
	Other Monthly Expenses Not Included in Exercise 1–2	
E	-	$
	-	$
	-	$
F	**TOTAL PERSONAL MONTHLY EXPENSES** (add rows D + E)	$
G	**PERSONAL CASH-FLOW RESULTS** Subtract expenses from your income (Row C – Row F)	$
H	Total Booth-Rental Expenses from Exercise 1–1	$
I	**TOTAL MONTHLY CASH-FLOW SUMMARY** (Row G – Row H)	$

In order to create our daily income goal, we are going to use this simple formula:

(annual income goal/50 weeks)/(No. of days worked per week)

This formula takes into account that you will work 50 weeks out of the 52-week year, taking into consideration that you will take two weeks of vacation throughout the year.

Let us use $50,000 as an example.

If your annual income goal is $50,000.00 a year and you work five days a week, your formula would be ($50,000/50)/5, which means $1,000/5 equals $200. The result means that in order to make $50,000 a year, you will need to make $1,000 a week and $200 a day.

Annual Income Goal	No. of Weeks Worked in a Year	No. of Days Worked in a Week
$50,000.00	50 weeks	5 days
	EQUATION = ($50,000/50)/5	
Weekly revenue required to make $50,000 a year	$50,000/50 = $1,000.00	
Daily revenue required to make $50,000 a year	$1,000/5 = $200.00	

If you focus on the daily goal, you can go into the day with your income goal in mind to help you achieve your annual goal. So whatever you desire to make in a year, it is recommended that you use salon software for independent beauty professionals or a recordkeeping appointment book to keep track of your income daily.

FIGURE 4–4
Establish Your Financial Dream Team.

4.4 Establish Your Financial Dream Team

Having a team of professionals who specialize in helping you keep your finances in order is a must in running your booth-rental business (Figure 4–4).

This is another area that is neglected by booth renters. Maintaining the finances of your business is very crucial because if this area is mismanaged, there can be serious consequences, such as being audited by the Internal Revenue Service (IRS), and no one wants that.

To keep your business finances in order, it is recommended you hire and build relationships with the following team of financial professionals:

Bookkeeper

It is a requirement of every business owner to keep a good financial record of their books. This is important because it is a requirement of the IRS that any revenue generated by the business be accurately reported. There are several software programs available on the market where you can keep track of this information yourself, but it is recommended for all businesses to hire a bookkeeper. A **bookkeeper** is a person skilled in the area of accounting that keeps track and records the financial transactions of a business. A bookkeeper will receive documents from you on a monthly or quarterly basis such as cash receipts, credit-card statements, bank statements, merchant statements, and payroll statements. Upon receipt of these documents, your bookkeeper will record all expenses and purchases made by the company, as well as record all revenues generated by the company in order to produce your company's financial statements such as a balance sheet and profit and loss (P&L) statements *(we will cover these statements in Section 4.5, "Analyzing Your Financial Health")*. When searching for a bookkeeper, keep in mind that it is important to find a bookkeeper who understands your industry. This is significant because you want to make sure that the bookkeeper categorically understands your purchases and expenses as it relates to your business.

Accountant

An **accountant** is a person who "audits and inspects the financial records of your business and prepares financial and tax reports."[1] An accountant will assist you with tax-planning strategies to reduce the amount of your taxable income, help lower your tax rate, declare available tax credits, help you keep track of when taxes are to be paid and help you to get the most out of your business expenses.

Financial Planner

A **financial planner** is a skilled individual who assists individuals and businesses achieve their long-term financial goals by analyzing

---

FYI

To actively participate in understanding how to organize your financial records and your financial accountability, take an accounting class at a local technical college. Learning the basics will help you understand what your bookkeeper and accountant are recommending.

the client's current financials and designing a plan to help the client achieve their financial goals.[2] A financial planner is a necessity in the growth and development of your business. A financial planner will help you with decisions as they relate to investing, insurance, and retirement.

Personal Banker

A **personal banker** is an employee of a financial institution who helps clients manage their assets, such as mortgages and savings, checking, and money-market accounts. A personal banker is a great asset to your financial dream team because they can help you with different products and services that a bank has to offer on a one-on-one basis. Personal bankers are also available to help you with filling out loan applications and the requirements needed to get the loan approved. This relationship is a great one to have, especially as your business grows, because your banking needs will change, and having a personal banker to communicate your business goals and needs makes the growth and development process a little bit easier.

Insurance Agent

An **insurance agent** is a representative of a state board–licensed insurance agency that sells insurance policies to customers. As the final member of your financial dream team, an insurance agent is a great asset to help you with your insurance needs. As we discussed in Chapter 1, it is important for an independent beauty professional to have personal liability and disability insurance, as well as other kinds. An insurance agent will look at your financial environment to determine how much insurance is needed to protect your assets.

Hiring a bookkeeper or accountant will require fees for services rendered. If your budget does not allow for these services at this time, there are software programs you can purchase that will allow you to do your own bookkeeping and accounting, such as QuickBooks by Intuit. If you can afford these professionals, it is recommended that as you build your financial dream team, you interview several candidates and also solicit referrals from coworkers and other industry professionals before you make your final selections. Remember, regardless of who is on your team, it is your money and thus ultimately your responsibility. You should never turn over the handling of your finances blindly. Take the time to understand what your team is recommending, why they are recommending it, and how you can improve your finances.

4.5 Create a Financial Health Checklist

As a small business owner who is responsible for every facet of your business, such as providing hair services and running the back-door operations of your business, things can easily get out of hand with the financial side of your business. One way to get a handle on your finances is to create a financial checklist (Figure 4–5). Opening a business checking account in your business name is another way to get a handle on your finances. Opening a separate business account will keep business finances separated from personal finances. All monies collected for services rendered and retail sales should be deposited into this account. This way, you can definitely keep your personal finances separate from your business finances.

Depending on your business structure, you will need your articles of incorporation or organization, along with your tax identification number to open the business bank account. Once your business account has been established, it is time to create your financial checklists. As a booth renter, you should have five different types of financial checklists:

1. Daily Checklist
2. Weekly Checklist
3. Monthly Checklist
4. Quarterly Checklist
5. End-of-Year Checklist

FIGURE 4–5

Create a Financial Checklist for Your Business.

Daily Financial Checklist

- ☐ **Cash on hand:** You should know and record your starting cash amount at the beginning of the day.
- ☐ **Service revenue:** At the end of each work day, summarize and record the total amount of revenue earned in services.
- ☐ **Retail revenue:** At the end of each work day, summarize and record the total amount of revenue received in retail sales.
- ☐ **Tips:** Record the amount of tips received for the day.

☐ **Credit-card terminal batching (if applicable):** Many credit-card companies require that a daily batch out be done on the credit-card terminal at the end of the day to ensure credit-card payments received are deposited into your business bank account in a timely manner.

☐ **Bank deposit totals:** Fill out or create your deposit statement for daily cash, check, and credit-card bank deposits.

Weekly Financial Checklist

☐ **Payroll recordings:** Payroll recordings should include the name of the employee (including yourself), the rate of pay, number of hours worked, the pay period ending date, and tax withholdings, if applicable. Yes, you should pay yourself as if you were an employee. Just taking money when you need it will only cause confusion and an inaccurate reporting of your income.

☐ **Booth-rental payments:** Many booth renters pay their booth rent on a weekly basis. Set aside your rent for payment during this time. Be sure to collect a receipt of payment.

Monthly Financial Checklist

☐ **Retail sales tax report:** This report should be printed every month to determine your local sales-and-use tax payment to your state government for taxes collected on retail goods sold.

☐ **Bank statement reconciliation:** This process compares the bank's record of deposits match with your records. You also will want to make sure that the daily credit-card batch totals match with your bank's deposit total.

☐ **Bookkeeping reconciliation:** Gather all your bank statements, credit-card statements, and receipts from the previous month and send them to your bookkeeper for accurate recordkeeping.

☐ **Balance sheets:** Receive a balance sheet along with a P&L statement from your accountant. Review each and ensure that you understand what you are reading. Schedule regular meetings to review, and ask any questions that you may have.

☐ **Account payables:** Pay all accounts due for payments.

☐ **Inventory:** Take inventory to determine what products you need to purchase and to determine what products did or did not sell so you can determine what promotions you should run or what discounts need to be given.

Quarterly Financial Checklist

☐ **Tax payments:** Make quarterly self-employment tax payments. Paying your employment taxes quarterly is a great way to keep your business on track financially. Many self-employed professionals who wait until the end of the year can find themselves overwhelmed with the amount of the payment due for taxes for the entire year. This leads to making monthly payment arrangements with the IRS, which is charged with interest. Contact an accountant to help you set up quarterly tax payments.

End-of-Year Financial Checklist

☐ Prepare tax documents such as W2s or 1099-MISC forms for employees or independent contractors who worked with you during the year.

☐ Take a look at your balance sheet, P&L statements, and income sheet to determine if your company made money for the year. After careful review of these statements, make the necessary adjustments to your business so it can be more profitable in the upcoming year.

☐ Take a look at your short-term goals to determine if you met them. If you have not yet, use this time to describe why those goals were not met and what you need to do to meet them.

☐ Complete an inventory. Take a look at what products you have in inventory at the end of the year. Create an inventory report that will show the value of inventory you have left at the end of the year. This information will be requested by your accountant for tax purposes.

Staying on track with your financial checklist throughout the year will really help you keep financially organized. Hiring an accountant and a bookkeeper is the key to success to help you manage your financial checklist successfully.

4.6 Analyze Your Financial Health

Getting an annual checkup on your financial health is just as important to your business as having an annual checkup is on your physical being. It is important to know where your company stands financially. In order to know what your company's financial health

---

RESOURCES

- Milady's Financial Analysis and Coaching Tools is an easy, step-by-step workbook, with audio tutorials and Excel spreadsheets, all designed to help you track your finances and make the most of your money. Features include nine spreadsheets to track, analyze, and record your financial numbers. Spreadsheets such as an inventory analysis, weekly–monthly tracking, an income-potential spreadsheet, and a pricing-your-service-for-profit spreadsheet are just to name a few. This powerful business tool also includes a Microsoft Excel basics tutorial, a PDF template of a QuickBooks chart of accounts, and an accompanying workbook offering suggestions and directions to implement the spreadsheets. You can find more information at this website: http://www.milady.cengage.com/business/financialanalysisandcoaching-tools.html

is, your bookkeeper will create several financial statements based on your revenue and expenses that will show you exactly how your company is doing financially.

The two primary financial statements that your bookkeeper will produce from the revenue you generated and your company's expenses are an income statement and a balance sheet (Figure 4–6). An **income statement**, also known as a **profit and loss statement**, explains all the income and expenses generated by the business over a period of time and depicts how profitable a company is. The **balance sheet** is a picture of your company's financial situation over a certain period of time.[3] Your company's balance sheet will show three things: the company's assets, its liabilities and net worth, and its equity. Let us explore these three areas of a balance sheet.

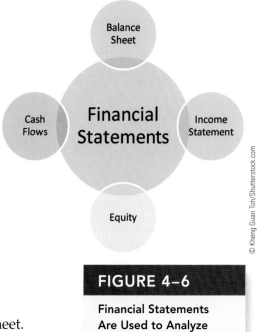

FIGURE 4–6

Financial Statements Are Used to Analyze Your Business's Financial Health.

Assets

An **asset** is defined as anything of worth on a balance sheet that is owned or payable to the company. Assets can be classified into three different categories: current assets, fixed assets, and intangibles.[4]

Current Assets

Current assets are assets owned by the company that are expected to be converted into cash in less than one year.[5] Current assets include the following:

- **Cash:** Cash includes all monies in the bank, such as in checking, savings, certificate of deposit (CD), and money market accounts.
- **Accounts receivables:** Accounts receivables are monies due from customers. These occur when a business sells inventory or services to a customer on terms, and monies are not collected at the time the goods or services are delivered. The customer is sent an invoice for payment of goods to be paid at a later agreed-upon date.
- **Inventory:** These are the products that are purchased by the company to be resold at a profit.
- **Prepaid expenses:** Prepaid expenses are payments made for goods or services that will be received at a later date (such as gift certificates).

Fixed Assets

Fixed assets are defined as monies used to buy physical assets, which are used to produce income and which will not be turned into cash within a year's time. Fixed assets are considered to be things such as real estate, equipment, machinery, and furniture.

Intangible Assets

Intangible assets are defined as monies used to purchase an asset that may never materialize into cash and has an undetermined life expectancy. Examples of intangible assets are patents, trademarks, and formulations.

Liabilities and Net Worth

Liabilities are defined as monies on a balance sheet that the company owes to creditors, such as loans, credit-card debt, and mortgages. **Net worth** is defined as the value of a company where the company's assets exceed the company's liabilities.[6] The goal of any company is to decrease its liabilities, which will increase its net worth.

Equity

Equity is defined as the amount of money contributed by the owners and the amount of ownership one has in the company after all the debts have been paid off.

Analyzing your company's financial health is crucial to the success of your business. As you can see, having a financial statement prepared by an accountant reveals how much money your company actually made after all the expenses were paid. Hiring an accountant to help you create a quarterly P&L statement is key to running a financially successful booth-rental operation.

4.7 Health Benefits

Maintaining your overall health is crucial to the success of your business because, without you, the business does not make money. As an independent beauty professional, the salon from which you rent will not offer you health benefits. A salon can only offer health benefits by law to those who are employed by the salon; therefore, it is important you seek out an individual health benefit plan (Figure 4–7).

An individual health benefit plan is health insurance that is obtained independently by an individual and is not affiliated with the benefits of the workplace. Many of the major medical insurance companies offer individual health benefit plans. It is recommended that you talk with

> **HERE'S A TIP**
> There are several industry associations that, as part of their member benefits, offer group rates for health insurance, liability, and so on. Examples include http://www.probeauty.org and http://www.insuringstyle.com.

your financial planner or insurance agent to point you in the right direction to find the best plan that fits your needs.

If you are married and your spouse has health insurance through their workplace, it is recommended to be added to your spouse's health benefit plan. If you are not married or do not have that option, you can become a member of a beauty industry organization that offers health benefit plans to their members.

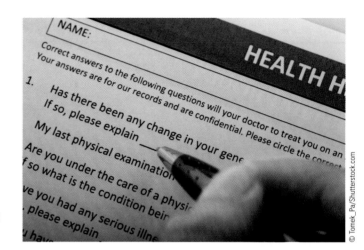

FIGURE 4–7

Obtain an Individual Health Plan.

4.8 Taxes

Paying taxes on revenue earned from your booth-rental business is a requirement of the IRS (Figure 4–8). It is also a requirement as a self-employed individual if you want to purchase personal items such as houses and cars. For example, if you desire to purchase a home, a self-employed individual is required to present their last three years' tax returns to the mortgage lender as part of the approval process. This is important to mention because there has been a trend amongst booth renters for years where they perform services on clients but did not report the income. Many of them were making a decent wage but could not show how they were making this wage when it came to making purchases like homes and cars because the income was not reported properly. Not only is this habit against the law, but it is damaging to your personal credit rating, and it reflects poorly on the professional beauty industry as a whole.

FIGURE 4–8

Paying Taxes on Revenue Earned Is a Requirement.

After servicing your clients, you will receive payments from your customers for services rendered in several forms, such as cash, credit cards, and checks. It is a requirement of the IRS that all revenue generated by the business be reported, including monies received for services rendered, retail products sold, and tips received from customers. To assist you with the accurate daily tracking of your income, it is recommended that you purchase salon software for booth renters that will allow you to record and keep track of client transactions

COACHING NOTE

Every salon/spa professional is required by federal law to report all income, including tips, as taxable wages. Tips are taxable wages: You are required not only to report them as income but also to pay the required FICA taxes on those tips. This is a *required* law; compliance with it is not optional. Underreporting income devalues you as a professional and the industry as a whole. Failure to accurately report all income has consequences.

and revenue. This type of software or app is a great income-management tool that will help you provide the necessary financial information to your accountant so a financial statement can be generated.

Now do not be afraid of Uncle Sam, because being a small business owner has some great tax advantages. Most of the things you purchase for your business can be used as a tax write-off to help lower your taxable income.

When you spend money on items to run your business, these items are called expenses and should be categorized for tax purposes. Your accountant will also ask you to categorize your expenses, so let us explore these expense categories further:

Advertising

You can write off any money spent on advertising such as flyers, postcards, billboards, radio and newspaper ads, television commercials, photo shoots for advertising in hair magazines, and any other type of advertising purchases used to promote your business.

Mileage

The miles you put on your car for business, such as driving to purchase supplies, driving to trade shows and classes, and driving to any event pertaining to your business, can all be written off as an expense (Figure 4–9). You must, however, keep track of all the miles you expense by recording the mileage driven, the date, and the purpose or event for the miles. There are mileage tracker apps available for your smartphone or tablet that can be used to record your mileage to keep an accurate record for your expenses.

Insurance

The fees you pay for your professional general liability insurance can also be written off as an expense.

Interest

Did you acquire a line of credit or business loan to build your independent beauty business? Did you acquire a business credit card to make purchases for your business where you are paying

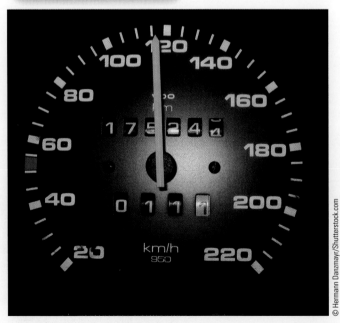

FIGURE 4–9

Keep Track of Mileage Driven for Business for Tax Purposes.

interest (Figure 4–10)? If you answered yes, then file away your monthly statements and turn them in to your accountant so the interest paid can be written off as an expense.

Office Expenses

Items purchased for office/business use, such as towel services, business cards, hair magazine purchases, magazine subscriptions, tickets purchased for hair classes, stamps/envelopes purchased for mailing promotions, appointment books, pens, pencils, professional memberships, and so on, are tax-deductible and should be categorized as office expenses.

Rent or Lease

The booth rent you pay to the salon can be written off and expensed under the rent category (Figure 4–11).

Repairs and Maintenance

Are you responsible for the upkeep of your area? Any monies paid to repair equipment or maintain your area can be written off in this area. This mostly applies to an independent stylist who leases a mini-salon at a salon suites complex. Any repairs that have to be made to your mini-salon for upkeep and cleaning can be written off.

FIGURE 4–10

File Your Monthly Statements. Interests Paid On Loans Are Tax-Deductible.

FIGURE 4–11

Payments for Booth Rent Are Tax-Deductible.

HERE'S A TIP

You can use the Repairs and Maintenance tax category to write off money spent to sharpen your shears.

FIGURE 4-12

Salon Supplies Purchased Are Tax-Deductible.

Supplies

This is a huge category for all booth renters. Every dollar spent on products such as color, perms, relaxer, shampoo, conditioner, combs, brushes, hair spray, blow dryers, rollers, gels, shears, hair irons, and so on can be written off under this category (Figure 4-12). Stylists spend more money on supplies than on any other category.

Taxes and Licenses

The fees you paid to renew your cosmetology and business licenses are also tax-deductible and should be written off in this category.

Travel

Any plane, bus, train tickets, or hotel costs spent for business travel such as going to a beauty trade show should be expensed in this category.

Meals and Entertainment

Any meals purchased while on business travel or if you are taking someone to lunch to discuss business will be expensed in this category. There are some great tax advantages you have as an independent hairstylist, so make sure you contact an accountant for more information about filing and categorizing your tax-deductible expense. Remember for everything you write off, please keep the receipt in case of an audit.

> **HERE'S A TIP**
> When purchasing items for your business, it is recommended you make all purchases via credit card or check to use as a tracking method for purchases. It is not recommended you purchase items for your business with cash, but if you do, keep the receipts for all cash purchases.

4.9 Saving for Retirement

Thinking about retirement is one of the last things people think about when talking about how much is needed to succeed in business. It should, however, be included in the equation of success when starting your business. One realization you must face is that one day you will get older and your body will give you signs that it is time to lay down the shears and retire (Figure 4-13).

As an independent beauty professional, you are responsible for saving and planning for your own retirement. When saving for retirement, self-employed individuals can save money monthly in a Simplified Employee Pension Individual Retirement Account, also known as a SEP IRA. The money you contribute to your SEP IRA is tax-deductible. For more information on setting up an Individual

Retirement Account, contact a financial planner to help you with this process.

Many people have questions about the right time to start saving for retirement. The answer to that question is, as soon as you start your career you should start saving for retirement. The younger a person is when they start saving for retirement, the longer you have for your money to mature. Let us explore how your money can mature depending on when you start saving for retirement.

FIGURE 4–13

Plan Ahead and Save for Retirement.

Example 1

Steve is 25 years old and is saving $250.00 a month, which is equivalent to $3,000 a year, in a tax-deferred retirement account that has an 8 percent annual return. Steve contributes to his retirement account for a period of 10 years and then decides to stop making deposits into his retirement account. Steve has contributed a total of $30,000 over a 10-year period. When Steve reaches the age of 65, his $30,000 investment with an 8 percent annual return has now grown to more than $472,000 over 40 years, and Steve did not contribute any more money to his account beyond the age of 35.[7]

Example 2

Steve does not start saving for retirement until he is 35 years old. He saves $250 a month ($3,000 a year) for 30 years in an account that has the same 8 percent annual return. At the end of 30 years, Steve would have saved $90,000 and the 8 percent annual return would grow this investment to $367,000 a year.[8]

In the two examples given, you can see the importance of starting early to save for retirement.

4.10 Build Business Credit

Many people when starting a new business use their personal credit cards to purchase items needed to get their business up and running. The problem with using your personal credit cards is that the debt accrued from using your personal credit cards is reported to your personal credit file as personal debt, and this pulls on your personal credit score. When your personal credit score is compromised because

FIGURE 4–14

Building Credit for Your Business Must Start with Setting Up Your Business Correctly.

of business purchases, you create financial challenges personally. In order to keep your personal finances separate from your business finances, it is recommended that you start building credit for your business that is separate from your personal credit (Figure 4–14).

As discussed in Chapter 2, the first step in building business credit for your business is to structure your business as a limited liability company, S corporation, or C corporation. If you are currently operating your business as a sole proprietorship and want to start building business credit, restructuring your business as a corporation is the first step. By incorporating your business, you can apply for your business credit without using your personal credit information to determine your approval. Your approval for a business credit line will be based only on your tax identification and corporate information. If you do not incorporate your business, you can still build a business credit profile, but you will be responsible for all the debts of the business because, as a sole proprietor, you must personally guarantee everything. If you operate your business as a sole proprietor or partnership, when you apply for business credit your personal information will be used to determine whether you receive the credit or not. So it is recommended you build your business credit as a corporation or limited liability company.

When building business credit for your business, it is important to know that there are some things your company must have in place in order to be successful in the business credit building process:

- Your company must be incorporated.
- Your business must have a business telephone service listed in the yellow pages.
- Your company should have a Web site.
- Your company should have a profile listed with Dun & Bradstreet, also known as D&B, which is a business credit bureau reporting service.

The advantage of building business credit is that you can separate your personal credit file from your business credit file, and purchases made on your business credit cards will not show on your personal credit profile.

It generally takes about two years to build credit for your business, but just like with personal credit, when you start building your business credit, your business creditors will report your payment history to business credit reporting bureaus. The top credit bureau for business is Dun & Bradstreet. Other business credit bureaus are Experian Business, Equifax Business, and Business Credit USA, to name a few. So once you build your business credit, it is important to pay on time to receive a good score, because just like with personal credit, you will obtain credit based on your score once your profile is built.

4.11 Summary

As you have learned in this chapter, succeeding financially as a booth renter is dependent mainly upon your personal financial obligations. Starting a booth-rental business does not require a lot of overhead costs. In order to succeed, you need to make certain you build a solid business structure and make sure you analyze your personal finances along with your current source of revenue before you decide to transition to a booth-rental operation. Let us review some valuable information you learned in this chapter. Use the following top takeaways to confirm the five steps it will take for you to succeed.

4.12 Top Takeaways: Your Money, Your Future

This chapter was filled with plenty of information to give you insight on how your finances play an important role in the success of your booth-rental business. Let us explore some top takeaways from this chapter.

- **Revenue.** Creating positive cash flow through generating more revenue than expenses is necessary in running a successful booth-rental business.

- **Income goals.** Setting an annual income goal will help you focus on your revenue-generation efforts. Once you set your annual income goal, break your goal down to a daily goal to give you a more attainable goal to focus on.

- **Recordkeeping.** Keeping accurate financial records for your booth-rental business is important so as to insure proper reporting to the IRS. Hire an accountant, bookkeeper, or use bookkeeping software to help you keep track of your revenue and expenses.

- **Saving for retirement.** Planning for retirement should be included as part of your expenses. Using part of the tips you generate to save for retirement is one method to secure your retirement future without pulling monies from your main income.

- **Build credit.** If you want to build credit for your business so that you can obtain credit in your business name instead of using your personal credit for purchases, it is recommended that you structure your business as a limited liability corporation, S corporation, or C corporation. If you are operating as a sole proprietor, contact your accounting professional to change your business structure.

---

[1] Accountant. (n.d.). The American Heritage® Dictionary of the English Language, Fourth Edition. (2003). Retrieved from http://www.thefreedictionary.com.
[2] Financial planner. (n.d.). Investopedia. Retrieved from http://www.investopedia.com.
[3] Preparing Financial Statements. (n.d.). The U.S. Small Business Administration. Retrieved from http://www.sba.gov.
[4] Ibid.
[5] Current assets. (n.d.). Investopedia. Retrieved from http://www.investopedia.com.
[6] Networth. (n.d.). Investopedia. Retrieved from http://www.investopedia.com.
[7] Retirement basics: Getting started – Ultimate guide to retirement. (n.d.). CNNMoney - Business, Financial, and Personal Finance News. Retrieved from http://money.cnn.com.
[8] Ibid.

Chapter 4 Quiz: Your Money, Your Future

This chapter describes how to handle your finances with effective money management and financial planning. Answer the following questions to review what you have learned.

1. List at least five categories that you can use for tax write-offs.
    1. _____
    2. _____
    3. _____
    4. _____
    5. _____

2. List five things your business success will depend on.
    1. _____
    2. _____
    3. _____
    4. _____
    5. _____

3. List three members you should have on your financial dream team.
    1. _____
    2. _____
    3. _____

4. You can write off the miles driven to and from any hair trade show or beauty-supply store.
    a. True
    b. False

5. What is the short name of the retirement fund that you can set up for yourself as an independent contractor?
    a. SEP account
    b. SET account
    c. SIP account
    d. None of the above

6. What tax category would you use to write off fees paid to renew your cosmetology certificates?
    a. Advertising
    b. Insurance
    c. Taxes and licenses
    d. All of the above

7. As an independent contractor, which of the following are supplies you can write off for tax purposes?
   a. Shampoo, conditioners, and haircolor
   b. Blow dryers and curling irons
   c. Combs and brushes
   d. All of the above

8. You must keep receipts for everything you write off for taxes, except your cell phone bill.
   a. True
   b. False

chapter 5

Marketing Your Booth-Rental Business

CHAPTER OUTLINE

5.1 Understanding Your Brand

5.2 Creating a Marketing Plan and Marketing Strategy

5.3 Marketing and Promotion Strategies

5.4 The Marketing and Promotions Calendar Template

5.5 Summary

5.6 Top Takeaways: Marketing Your Booth-Rental Business

Career Profile

Dawn Medina

Dawn Medina has been a booth renter since 2008. Going independent has allowed her the flexibility to cater to her clients' needs and schedules. This business structure lets her build a business case and develop her vision so she can one day open her own salon. Since becoming an independent hair stylist, she has implemented several marketing and self-promotion programs, such as monthly specials, client loyalty programs, and a personal Facebook (FB) page.

How did you become an independent stylist?

In March 2008, I made the transition into booth renter as a sole proprietor. The down economy had forced the current salon owner to make the switch to booth renting. I had mixed feelings and concerns on how to set up to become a sole proprietor. My independence came naturally. My strong professional work ethic and ability to implement my own business programs helped win me great success! Having monthly specials, a client-loyalty program, and a client-referral program, as well as my choice of retail products, were a few changes I made. I am currently at the end of my fifth year as an independent stylist and have been blessed with steady financial growth and client retention.

How do you market your independent beauty business? Do you have a favorite promotion strategy in your past that worked really well?

I have used flyers around the neighborhoods and ad spots on the placemats of area restaurants to attract new clients. I have a good client base already built up, so I keep my name on their minds by using a client e-mail list, and I do this by offering specials every month. I ask everyone if they would like to be added to the e-mail list to receive the monthly specials. I also offer occasional online-only specials during slow times to generate more business and notify customers of schedule changes. In addition, I use Facebook (FB) and have it all connected through my smartphone. Clients are able to see promotions and schedule changes, view new product info, etc. I have found a design that reflects who I am and I use it in all of my advertising materials. I used to do "cut club cards," after 10 full-price haircuts, get the 11th free. I changed it this year to after ten salon services, get $10 off of salon services or my retail products. FB is also nice for this because my clients tag me in their posts or pictures and then their friends message me for info on wanting to schedule with me. Once they are added, they can view a lot of my work on my page; it is free advertising and spreads faster than anything. Once a new client is referred, I send a thank-you postcard out to the client who made the referral. It is worth $10 for them to use with me, and I do this every time. I print my own flyers. My cards and postcards are done through 123 Print, an online printing service, and they carry a large variety of matching low-cost material.

Is there anything you would like to share with future booth renters about creating your brand, creating a marketing plan and a marketing strategy, and the successful or unsuccessful marketing and promotion strategies you have used?

Talk to people who have or are doing it. Find out what worked or did not work for them. Take the time to call around and price the material or service you need to market your business. Stay current; I have found the social networks are accessed by nearly everyone, and that is FREE! Keep your name on others' minds. I get so much positive feedback from clients who love the incentives that it can sometimes influence them to come in now rather than next week. I work with four other women who are all booth renting. I am the only one who advertises and offers what I do to the clients. IT DOES MATTER! I have the highest volume of business coming in (even over the shop owner). It has attracted the attention of others' clients. If it is slow, get on FB and advertise a special for that day only on haircuts or color!

Marketing your independent beauty business is vital to the success of your business. There are several ways to market your business to increase clientele. In this chapter, we will cover the different ways you can market your independent beauty business.

5.1 Understanding Your Brand

When marketing your business, the first thing you must understand before you go to market is your brand. A **brand** is defined as a distinctive graphic design, stylized name, unique symbol, or other device used in an image, logo, mark, phrase, or combination of these that businesses use to differentiate themselves from other businesses in the market.[1] A brand is the essence of what the business is and aligns with the vision and values of the company.

Creating an image, logo, or tagline for your business is beneficial to bringing brand awareness to your company. The decision to create a logo with an image, a symbol, or just text is completely up to you (Figure 5–1). A **logo** is defined by BusinessDictionary.com as "a recognizable and distinctive graphic design, stylized name, unique symbol or other device for identifying an organization."[2]

When you design a logo, make it recognizable and memorable. Coca-Cola, McDonald's, and Kentucky Fried Chicken are a few examples of this. Each of these brand's logos are immediately recognizable no matter where you go in the world. Coca-Cola's logo uses only graphic text, McDonald's employs the golden arches as their logo, and Kentucky Fried Chicken uses a photo of Colonel Sanders, along with the text "KFC" in their logo. Whatever you decide, it is recommended you use a logo design company or professional graphic artist to design your logo, because this will be the mark that represents your brand.

The goals of a good brand will:[3]

- Convey the message of the brand clearly
- Validate the brand's credibility
- Draw in the prospective buyer emotionally
- Inspire the consumer
- Build brand loyalty

Think about it. There are only certain brands that you will use because you

FIGURE 5–1

A Logo Can Be Created with Only an Image or Text!

are emotionally connected to the brand. Even if you come across a knockoff or similar product, do you typically continue to purchase your brand of choice? In most cases, the response is "yes" because you have tried and trusted a certain brand. During your brand development, you must clearly understand what message you want your brand to send when a customer sees it so that you can build brand loyalty among your consumers. Once you understand your brand, and what you want your brand to portray when it is seen in the market, you can effectively market your brand to consumers.

5.2 Creating a Marketing Plan and Marketing Strategy

Before you decide to launch a marketing campaign for your independent beauty business, it is recommended that you first create a marketing plan (Figure 5–2). A **marketing plan** explains your advertising and marketing goals for a calendar year.[4] It details specific actions you plan to take in order to influence customers to purchase the products and services you offer.

A good marketing plan will include seven sections:

1. **Introduction:** The introduction is a short paragraph describing the products and services you offer.
2. **Situation Analysis:** The situation analysis describes the current state of your business and takes a look at your company's strengths, weaknesses, opportunities, and threats as they relate to marketing your business. A situation analysis will also include a S.W.O.T. analysis (see Chapter 3, Section 3.6: "S.W.O.T. Analysis").
3. **Target Market:** In this section, you will describe who your target market is. Describe the type of client you want to market your services to. Give a detailed demographic, such as gender, age, ethnicity, and profession.
4. **Marketing Goals:** In this section, you will write down your marketing goals for the year. List your short- and long-term marketing goals, for example, "Increase new clients at a rate of five per month," "Grow Facebook (FB) fans to 2000," and so on. Be realistic about your goals because you will have to come

FIGURE 5–2
Create a Marketing Plan Before You Start Your Marketing Campaign

back and evaluate how you did on reaching these goals.

5. **Marketing Strategy and Approach:** A marketing strategy supplies the goals and tactics for your marketing plans.[5] The marketing strategy section is a detailed account of how you will strategically approach and execute your marketing goals (Figure 5–3). This section is the largest section of your marketing plan, so be precise. For example, if your annual goal is to increase your clientele by 30 percent, you will describe how you plan to achieve this by detailing monthly or quarterly tools and techniques that will be used to reach this goal.

FIGURE 5–3

The Marketing Strategy Details How You Will Execute Your Marketing Goals

6. **Detailed Budget:** The detailed budget will itemize the cost of each marketing approach you intend to execute in your marketing strategy. For example, if you plan to launch a FB advertising campaign for six months of the calendar year, you would break down the costs associated with the FB advertising, such as designing the graphic, the purchase of any stock art needed to produce the ad, and the actual advertising cost.

7. **Evaluation:** Once your marketing plan is complete, you should track the results of each marketing endeavor. Doing so will allow you to analyze the effectiveness of your plan and determine how it is performing. When you evaluate each of your marketing strategies, your evaluation should ask the following questions:

- *Has your revenue increased?* Look and see how many new clients you have serviced since the implementation of your marketing plan. You should record all new clients and/or sales associated with each marketing initiative.

- *How did your new clients hear about you?* Whenever you service a new client, ask the client how they heard of you to determine if this client came in due to your active marketing plan. Record the results in your tracker.

- *Is your advertising campaign action-driven?* When producing your advertising campaign, make sure you invoke the customer to do something so you can measure your results. For instance, your marketing campaign should have the client call the business for a discount, or it should send the customer to your website to download a coupon that they will bring in to receive a special discount on a service. Using action-driven mechanisms will give you a true measure of the return on your advertising efforts.

- *Did your networking activities bring in new business?* One way to market your business is to participate in networking events where you will donate product samples or a certain amount of free services, all while spreading the word about your business. Remember, giving is a principle that helps you receive.

- *Does your marketing approach make it easier for you to sell your services?* When answering this question, evaluate whether your marketing approach attracts your target market, and evaluate whether the design of your print materials makes the customer want to keep your literature.

- *Are you getting a return on your investment?* Use the Advertising Tracking Form in Exhibit A at the end of the chapter to track all new business associated with each initiative. Evaluate whether the money you spent on your marketing plan is producing enough revenue to validate the costs associated with the campaign.

It is now time to create your marketing plan. Use Exercise 5–1 to develop your marketing plan and Exercise 5-2 to evaluate your marketing strategies.

MARKETING PLAN TEMPLATE

Use the marketing plan template in Exercise 5–1 to create your marketing plan.

EXERCISE 5–1   Marketing Plan Template

Section 1: Introduction (*Describe your products or services.*)

Section 2: Situation Analysis (*Complete a situation analysis. Use extra paper if necessary.*)

Section 3: Target Market (*List your target market.*)

Section 4: Marketing Goals (*List your marketing goals.*)

Section 5: Marketing Strategy (*Give a synopsis of your marketing strategy.*)

List your marketing approach or tactics.

1)

2)

3)

4)

5)

6)

7)

8)

(*Continued*)

EXERCISE 5–1 Marketing Plan Template (*Continued*)

Section 6: Detailed Budget *(List the cost breakdowns associated with each marketing approach.)*

Marketing Approach	Detailed Costs
1)	-
	-
	-
2)	-
	-
	-
3)	-
	-
	-
4)	-
	-
5)	-
	-
6)	-
	-
7)	-
	-
8)	-
	-

MARKETING STRATEGY EVALUATION FORM

Use Exercise 5–2 to evaluate the effectiveness of your marketing strategies.

EXERCISE 5–2 — Marketing Strategy Evaluation Form

Date: From _____ to _____

Has your revenue increased?
If yes, describe:

How did your new clients hear of you?

Is your advertising campaign action-driven?
If yes, describe:

Did your networking activities bring in new business?
If yes, describe:

Does your marketing approach make it easier for you to sell your services?
If yes, describe:

Are you getting a return on your investment?
If yes, describe:

According to your marketing evaluation, what marketing strategies were effective? Identify those strategies that worked and then continue them if desired. If you find that some strategies were ineffective, go back and tweak your plan and try a new approach.

Now that you have completed your marketing plan and have understood how to evaluate each strategy, let us talk about specific marketing and promotion strategies to grow your independent beauty business.

5.3 Marketing and Promotion Strategies

Participating in multiple marketing methods and offering promotions that are effective will determine the success of your marketing strategy. All marketing tactics and promotions are not effective in every business. For example, you would not want to advertise your business in a parenting magazine unless what you are selling is geared toward parenting. Conversely, offering a FREE color service with the purchase of a haircut is great if you do not have any clients, but it could financially bankrupt a seasoned stylist just to gain a few additional color services. In this section, we will discuss some marketing and promotion strategies that will return results for your booth-rental business.

Website and/or Online Portfolio

A website is one of the most popular ways to effectively market your independent beauty business. A growing trend in the beauty and wellness industry is to build online portfolios. Using companies such as http://www.styleseat.com or http://www.bloom.com, you can build an online portfolio showcasing your work and use many of the features found on a typical website. Having either a website or an online portfolio or both for your business is like having a business in two places: online and on-land (Figure 5–4). The only difference between the two is that your business online is available 24 hours a day, 365 days a year. Your actual physical storefront business has actual hours of operation. For an affordable annual fee, you could graphically advertise your business via the Internet with a website. Both your website and/or portfolio should display images of your hairstyling work, list the prices and services you offer, list the types of products you use, give biographical information about you, and give clients a look inside your business before they actually come in to receive a service. Let them know that additional services, such as online booking,

FIGURE 5–4

Having a Website Is a Cost-Effective Way to Market Your Business

are available, and offer a nice perk to new and existing clients. Having a website or online portfolio has proven to be a great addition to any business, and it is a surefire way for you to gain clients.

To create a website, it is recommended you use online templates or contact a professional Web designer to help you with your Web design. Once your website is complete, you will have to promote the website. One of the best ways to do this is to include your website address on all correspondence issued from your business. This includes any literature, marketing materials, or business cards you give out.

E-mail Marketing

E-mail marketing is a popular and inexpensive way to market your business (Figure 5–5). Sending e-mail blasts lets you send graphical or text promotions using e-mail addresses you have collected from your customers. There are several companies, such as Constant Contact (constantcontact.com) or Mail Chimp (mailchimp.com), that offer predesigned templates you can use and customize by adding your own images and company

FIGURE 5–5

E-mail Marketing—An Effective Tool to Market Your Business

> **HERE'S A TIP**
>
> Keep your company's brand in mind during the design of your website. Your website should incorporate your brand in various ways, such as by using color schemes, logos, marks, and brand taglines. Remember, your website is an extension of your brand.

information. Additionally, many salon software companies offer these types of services as upgrades. Be sure to check with your software company.

Text and Mobile Marketing

Text marketing is a marketing technique that is becoming very popular (Figure 5–6). Text marketing uses cellular phone technology to send out promotions and discount coupons to clients. Since the inception of smartphones, mobile marketing such as text marketing has been used by many businesses to send graphical advertisements and regular text promotions to its customers. Several companies offer this type of marketing at a low cost. For more information on text or mobile marketing, do a search on the Internet for text marketing companies.

Social Media Outlets

Using social media outlets like FB, Twitter, and YouTube is an excellent and a cost-effective way to promote your business (Figure 5–7). Social media outlets let you graphically and digitally promote your services or products to your friends, family, and other members of the social media world. For example, you can create a FB fanpage for your business that allows your clients to connect to your business. You can also use your FB fanpage to announce promotions, post photos, and give hair tips or any other information that can be used to further market your business. In order to keep your FB fans engaged in your business, it is recommended you plan to post to your page two to three times a week. The goal of making posts weekly is to keep your clients and fans engaged and informed yet not overwhelm your fans with too many posts.

You can also link your FB fanpage to your Twitter account so all the information you want to provide to your clients through social media is consistent. This also makes it easier for you to post all your information in one place.

It is beneficial to the success of your social media campaigns to include all social media icons and addresses on your website, print material, and business card. Today, many companies offer free and paid social media training. Often, if attending a hair show, you can find classes specifically focused on social media. Be sure to check out webinars offered by associations and companies such as the Professional Beauty Association, StyleSync, or Insuring Style.

FIGURE 5–6

Text Marketing—A Great Way to Offer Discounts to Customers

FIGURE 5–7

Using Social Media Outlets Is a Cost-Effective Way to Promote Your Business

Direct Mail

Direct mail marketing is another cost-effective way to market your business to potential customers. With direct mail marketing, you can advertise your business to new and existing residents who live within a certain distance (mile radius) of your business (Figure 5–8). Many direct mail service companies let you purchase the names of new residents who move into areas near your business. These names, also called the "New Movers List," can be purchased by a particular zip code, county, or city name. The name list can also be broken down demographically by age, gender, ethnicity, income, marital status, and dwelling type, such as a single-family home, apartment, or P.O. Box.

Once you have selected your name list criteria, the direct mail company will e-mail you a list of names generated according to your specific criteria in label format. Once the labels are received, you can send out nice graphically designed postcards to potential customers, offering discounts on first-time visits and any other specials you think appropriate.

FIGURE 5–8

Direct Mail Marketing Can Be Used to Promote Your Business to New Residents Who Move Into the Area

RESOURCES

Companies like Directmail.com and the United States Postal Service offer existing resident and new movers lists at a reasonable cost. For more information on these lists, visit directmail.com and usps.com.

Photo Shoots and Print Media

One fun and interesting way to market your independent beauty business is through participation in professional photo sessions. Advertising using professional photos can build your clientele and expose your work to a variety of potential new clients. Participating in professional photo shoots allows you to advertise, through print media, the different services and hair designs that you specialize in (Figure 5–9). The cost of participating in a professional photo shoot can cost anywhere from $150.00 to $200.00 per model, which includes professional makeup services. These expenses are tax-deductible. One way to curtail the expense of your shoot is to use your clients instead of professional models, or offer your services at a local fashion show, bridal, or prom event in exchange for professional photos of your work. You could also partner with a local photographer and offer your clients a glamour-shot photography session. The photographer is paid through packages that your clients purchase and the fact that you are bringing them new business. Your client will walk away with a professional glamor photography photo, and you can use the photo in your portfolio.

Participating in photo shoots helps you build a portfolio for yourself to send to recording studios, television networks, and hair product

FIGURE 5–9

Participating in Photo Shoots Allows You to Advertise Your Work through Print Media

companies. When you apply for such positions as platform artist, technical educator, or hairstylist for entertainment/television, one of the first things they will want to see is a professional portfolio.

Beauty magazines such as *Sophisticate's Hairstyle Guide*, *Hair Style*, *Short Hair*, and *Hype Hair* are always seeking new photos to display in their magazine. There is no charge to submit your photos to these magazines. Look in the inside of each of these magazines for details on how to submit and make sure the photo is eye-catching, and if you stick to their guidelines, you may very well see your work displayed in their next issue. Before any photo shoot is scheduled, make sure that all models/clients sign a model release form. Each magazine has their own model release form, so make sure you contact them to receive a copy before your models are photographed, because the hair magazine will not print your work without this signed form. Annual participation in this type of advertising strategy can prove to be beneficial in attracting new clientele.

Radio and Television Commercials

Advertising your business through television commercials via your local cable company or through your local radio stations can reach many potential customers. It is also not as expensive as you might think. The cost will vary depending on which cable or radio stations you choose to advertise on, which areas you choose to market yourself in, and how often your commercial is aired.

Example: Mary advertised twice a week on Lifetime TV with her local cable company. She advertised once every Sunday between the hours of 2 pm to 8 pm and once every Monday evening during Lifetime Movie Monday. Her average cost for the month was about $575.00. The cost to shoot the commercial was an additional $1,000.00. This advertising effort increased Mary's clientele by 20 percent.

Please note: *This is only an example of what one stylist did in her local market. Prices will vary.*

FIGURE 5–10

Billboard Advertising Is a Great Way to Generate New Clients

Billboards

Promoting your business through billboard advertising can also generate new clients (Figure 5–10). When using a billboard to advertise your business, it is important that the design of your billboard graphic be clear, concise, and make a brief but powerful statement that makes people want to receive the services you offer. Your billboard advertising should be photographical, give a call to action, and display your telephone or website information. The cost of promoting your business using a billboard can be more costly than the other mentioned marketing strategies, but you can find relatively cost-effective billboard advertising spots in local communities and on side streets located near your business. Billboard advertising is very attainable for an experienced booth renter.

FIGURE 5–11

Working as a Technical Educator Allows You to Market Yourself

Technical Educator/Platform Artist

Becoming a technical educator with a hair product company is another way to market yourself and your skills (Figure 5–11). When you work as a platform artist, you have the chance to travel around the world and demonstrate your work in front of audiences everywhere. This kind of attention is one of the best ways for people to see you perform in front of a large crowd, listen to your verbal skills, and also get a little taste of your personality. Becoming a technical educator will help build clientele because it is a great resume booster and can be used in describing your abilities as an independent beauty professional. Customers like to be serviced by independent beauty professionals who are in demand, such as a technical educator or celebrity stylist.

In preparing to become a platform artist, you should do the following things:

1. Perfect your styling techniques, verbal communication, and knowledge of the hair product company you wish to work for.
2. Build your cosmetology resume and portfolio.
3. Participate in hair trade shows and take your portfolio with you to talk with the different product manufacturing companies you are interested in working for.

Promotional Offerings

With each marketing method, you have the opportunity to promote a product or service. The examples provided next are meant to serve as a guide and stimulate ideas. First, select the marketing method (for example, FB) and then select the promotional offering (examples shown next).

Retail Promotional Ideas

- **Offer free travel-size samples** of your latest products for first-time clients. These products can be put in a gift bag that contains their purchased products from the visit, along with a copy of your menu of services, your business card, and a postcard offering further incentives and discounts by following you on your social media platforms.
- **Create a buy-three-get-one-free promotion** where if a client purchases three products you recommended on that visit, they can get one additional product of the same or lesser value for free. This is a great way to get slow movers off of your shelves while insuring that your clients leave with the products you used on them that day.
- **Hand out retail frequent-buyer cards** so that upon each visit when a client purchases a product, you punch their card. After 10 punches, they get a free product.

Add-On Services

- **Ten percent off any additional service** recommended during that visit. This can be offered every Monday or the first Monday of the month as an example.
- **Service frequent-buyer card:** Each visit where a client purchases an additional service, you punch their card. After 10 punches, they get a free service.
- **Color me crazy:** First-time color clients will get 10 percent off of their color treatment, and another 5 to 10 percent of the proceeds will be donated to a charity. You can run this for a month and connect it to the charity of your choice. This will boost color sales and the opportunity for recurring color business thereafter.

Rebooking

- **Roll the dice and win:** Any client that rebooks their next visit with you on the spot will get a chance to roll two six-sided dice (a 2 to 12 percent discount potential) to receive a discount on their retail purchase that day. Purchase a small bowl and fill it with multicolored six-sided dice. When the client rebooks, they pick out two dice and roll them. You can even decorate your station with fuzzy dice. Make it fun and they will come back. This promotion increases retention and drives retail sales at the same time.
- **Rebook your next five visits** and get the fourth visit at half off. This will drive retention and help you control your book, guaranteeing repeat business.

Referrals

- **20/20 program:** Any client who refers a friend who comes in will get 20 percent off of their next visit. The new client will also get 20 percent off.
- **Refer three friends and get 50 percent off** your next visit.
- **Referral contest:** Offer your clients a chance to win something like a $250 gift card to a local department store such as Macy's or Nordstrom. Run the promotion for a month. Every referral card that comes in with the referring client's name on it gives them another chance to win. The more cards in the bowl, the better the chance they have to win. At the end of the month, pick one winner and post their photo and name on your social media outlets. Make it a big deal and people will play. You can offer gift cards for shopping or a dinner and a movie or a day trip with you somewhere fun. Get creative. You will more than recover the small investment of $250 by the end of the month and beyond, based on the number of referrals you get. What is critical is that you talk it up and make it fun for you and for them!

Rewards Program

Create a rewards program where your clients receive a rewards card that they can use to build points. This type of loyalty program can be similar to a miles program that airlines offer. You can offer point values for every dollar spent on services and products. Escalated point values can be assigned for an add-on service or for multiple products purchased per visit. You can offer double bonus point specials throughout the year in conjunction with holidays or seasonal trends, like back to school, their birthday month, and so on. If they achieve a certain point level in a calendar year, they can qualify for special rewards and prizes like mugs, T-shirts, and gift cards.

As you can see, there are several marketing and promotional strategies you can use with your marketing campaigns. Get creative, try what appeals to you, and track each promotion so you can analyze its effectiveness.

5.4 The Marketing and Promotions Calendar Template

Now that we have discussed the different types of marketing methods and promotional strategies that can be used to market your business, use the marketing and promotions calendar template in Exercise 5–3 to help guide your marketing strategies for the year.

EXERCISE 5-3 The Marketing and Promotions Calendar

	Marketing Method	Ad Size or Length of Promotion	Cost of Advertisement	Date of Launch
JAN			$	
FEB			$	
MAR			$	
APR			$	
MAY			$	
JUN			$	
JUL			$	
AUG			$	
SEP			$	
OCT			$	
NOV			$	
DEC			$	
Total Advertising Costs:			$	

5.5 Summary

As you can see, effective marketing is a bit more complicated than taking a flyer and putting it on someone's car or standing out in front of the local grocery store passing out business cards to generate business. It takes a viable plan to be put into action in order to know where you are going and whom you are trying to target.

Marketing your business is very important and should be ongoing to establish a loyal client base. In this chapter, we spent a lot of time talking about marketing to generate more clients, but it is equally important to retain the clients you are already servicing. Creating client partnerships is important for the success and growth of your booth-rental business, and will be discussed in the next chapter.

Let us review the top marketing takeaways to refresh your memory about the importance of marketing your business.

5.6 Top Takeaways: Marketing Your Booth-Rental Business

- **Branding.** Before you begin a marketing plan, you must understand your brand. Understanding your brand will allow you to choose a marketing plan that is suitable for your independent beauty business.

- **Logos.** During the process of developing your brand, select a logo or mark that will represent your brand. Remember, your logo or text design should be clear and concise, because this is the mark that will represent your brand in years to come.

- **Affordable marketing strategies.** Marketing your independent beauty business does not require a lot of money. There are very affordable marketing techniques that will allow you to build your clientele.

- **Budget.** Create a detailed budget that will give you an estimated cost to execute your marketing plan.

- **Promotions.** Create a marketing and promotions calendar to help guide your marketing strategies for the calendar year.

---

[1] Brand. (n.d.). Investopedia. Retrieved from: http://www.investopedia.com.
[2] Logo. (n.d.). BusinessDictionary.com. Retrieved from: http://www.businessdictionary.com.
[3] Lake, L. (n.d.). *What is branding and how important is it to your marketing strategy?* About.com Guide. Retrieved from: http://marketing.about.com.
[4] Marketing Plan. (n.d.). Small Business Dictionary. Retrieved from: http://www.entrepreneur.com.
[5] Marketing Strategy. (n.d.). Small Business Dictionary. Retrieved from: http://www.entrepreneur.com.

Chapter 5 Quiz: Marketing Your Booth-Rental Business

This chapter introduced several tools you can use to advertise and market your business. Answer the following questions to review what you have learned.

1. List three objectives of a good brand.

    1. _____
    2. _____
    3. _____

2. A marketing plan and a marketing strategy are the same.
    a. True
    b. False

3. Which of the following item(s) will explain your advertising and marketing goals for a calendar year?
    a. A business plan
    b. A marketing strategy
    c. A marketing plan
    d. None of the above

4. A good marketing plan will include (a) _____.
    a. detailed budget
    b. marketing goals
    c. marketing strategy
    d. all of the above

5. List four cost-effective ways to market your business.

    1. _____
    2. _____
    3. _____
    4. _____

6. You can use _____ to list the prices and services you offer, display photos of your hair styles, and book appointments.
    a. the newspaper
    b. your website
    c. radio and television
    d. all of the above

7. Participating in photo shoots will not help you _____.
   a. completely eliminate your competition
   b. build your clientele
   c. advertise your different services and specialty hair designs
   d. build your portfolio to send to recording studios, television networks, and hair product companies

8. Which technique is becoming one of the more popular ways to market your business?
   a. Direct mail campaigns
   b. Billboard advertising
   c. Text and mobile marketing
   d. Television and radio advertising

9. Which social media icons should you include on your website and print materials?
   a. Facebook
   b. Twitter
   c. YouTube
   d. All of the above

10. When creating a promotion for a product or service, what do you do first?
    a. Select the promotion offering and then the marketing method.
    b. Select the marketing method and then the promotional offering.

EXHIBIT A Advertising Tracking Form

Company: _____
Cost: $ _____
Dates: from _____ to _____

Date	Client Name	Services Received	Dollars Spent	Technician

Analyzing Your Advertisement!

Total number of new clients: _____

Advertisement cost: $ _____

Dollars spent by new clients: − _____

Total cost of advertisement: $ _____

(Take **Advertisement cost** and SUBTRACT **Dollars spent by new clients**)

chapter 6

Strategies for Retention and Building Clientele

CHAPTER OUTLINE

6.1 Client Partnerships

6.2 Establish Client Incentives

6.3 Client Retention Strategies

6.4 Coaching Corner: Business Ethics

6.5 Summary

6.6 Top Takeaways: Strategies for Retention and Building Clientele

Career Profile

Kristen Wanger

Kristen Wanger has been in the beauty industry since she was a teenager working at various salons and makeup counters at the local mall. She made the move toward becoming an independent stylist early in her career because she wanted the freedom of choosing her own products, making her own schedule, and having the ability to maximize her income. As a young stylist, she has built her clientele through being personable and putting her name out to everyone she meets. She is currently renting in a salon and working toward one day owning her own.

How did you begin your career as an independent stylist?

I went to beauty school and afterwards got my foot in the door at a prestigious local salon. I was an assistant for my first six months and then became a commission stylist. After about a year I decided it was time to become an independent booth renter. Before I made the official transfer to being a renter, I knew there were things I had to look over and discuss with the salon owner and, of course, the one person I have never made a big decision without, my dad.

What was your business/financial plan when you first started?

I had to look over the numbers I had been making the last few months to see if I was making enough to afford the rent, along with all of my other personal expenses. Once I figured out it was feasible for me, I made the switch to renter. My dad owns a business, so he was and still is a great business coach for me. I am very thankful to have not only a great dad, but a great dad who I can trust as someone to confide in about my business and the financial decisions that come with it!

What are the challenges in booth renting?

Since I have been a renter, I would say the things that have been the most challenging so far are how expensive it really is, and how on top of things you need to be. You need to make sure you can afford the color orders, and all of the other supplies you need! For me personally, I also like to make sure that I always have the extra funds to attend classes to better my skills. With that being said, I have also been able to triple my income within two years of being a renter.

What role did building clientele play in your early success?

Constantly bringing in new clients is a very important part of my business. The two main ways I have built clientele are simple: handing out my card to as many people as I can, and keeping people updated about my work through effective conversations and social media.

What are some strategies for maintaining your returning clients?

For all first-time clients, I send them personal thank-you cards with a discount offer card inside for their next service. This is a great way to get people to come back for a second visit. Asking clients a lot of questions and really getting to know them is always good practice, too. When clients come back in for a service, remember things that they told you and ask them about it. I think that the more thoughtful you are with your clients, and the more you get to know them and the more they get to know you, the more they will trust you.

Clients are the lifeline of any booth-rental business. Without clients, your business could not generate revenue.

Building a strong clientele is the key to running a successful booth-rental operation. However, building clientele is one of the biggest challenges that most booth renters face. It is important to understand that the client-building process is directly tied to the marketing of your business, which was covered in Chapter 5, "Marketing Your Booth-Rental Business." One age-old way to build clientele that does not cost any money is word of mouth. Word of mouth, or viral marketing, is still considered one of the most effective ways for a booth renter to build clientele. Referrals or recommendations are examples of word of mouth. It is the personal communication about a service or product between target clients and their friends, relatives, and associates.

Once you build a clientele, you want to focus on retaining your clients. The process of retaining clients has to do with building client partnerships where you become the client's preferred beauty provider. Client loyalty plays a major role in growing your business. Loyal clients will refer other customers, which results in additional revenue for your business. Good client retention is having 70 percent or more of the clients you service return to you for services in the same year.

6.1 Client Partnerships

Client partnerships are professional relationships built with clients to ensure customer satisfaction. In simple terms, it means you work to develop and nurture the relationship with your customer by communicating and providing incentives that will keep your customers engaged in your operations. Forming client partnerships is crucial to building your booth-rental business because client growth equals revenue growth (Figure 6–1). As you build your client partnerships, take a vested interest in your customer by finding out information from them such as their occupation and lifestyle habits. The key is to keep the conversation on and about your client.

FIGURE 6–1
Build Client Partnerships

Gathering client information helps you improve the services you offer by allowing you to cater your service to their individual needs and provide a wonderful one-of-a-kind experience that they deserve.

For example, if you have a client that participates in water aerobics, you can cater the service and product suggestions to fit the lifestyle of the client. Perhaps they mention that they are really sore, so providing an extra head, neck, and scalp massage says, "I care and I am listening." When you engage in these types of conversations and services with your customers, they will take notice and begin to refer other new customers. The referral of a new client from your existing client is one of the first ways to know that a successful client partnership is forming.

During the client-partnership development process, keep these things in mind:

- Client partnership is not a friendship.
- Client partnership is about focusing on the client, not on you.
- Communicate with your clients.
- Offer client incentives.

A Client Partnership Is Not a Friendship

Many booth renters confuse client partnerships with becoming friends with a customer. A friendship with your customer is totally different from a client partnership. Being friendly to your customer does not mean you are friends (Figure 6–2). As a booth renter, you will be in constant communication with your customer. Your customer will call you to schedule, reschedule, and cancel appointments. Sometimes, your customers may call you just to discuss a service they are thinking about trying. This continuous communication can sometimes be mistaken for a friendship that is forming instead of a partnership.

The one sign that the client relationship is changing to more of a friendship is that your customer will start inviting you to personal functions outside of the salon. You may be wondering why the conversation has channeled into this direction. The reason is that many booth renters find themselves in a situation where they have started on a successful path of forming client partnerships but the partnerships turn into friendships. Once the customer becomes a friend, then expectations from the customer begin to change as well. This is a common flaw in the client-partnership building process between booth renters and their customers. There are many scenarios where the hairdresser becomes comfortable about showing up late or not showing up at all for scheduled

FIGURE 6–2

There Is a Difference Between Client Partnerships and Friendships.

appointments with their customer due to expectations from this new relationship. Clients have also been guilty of taking advantage of the new relationship by not showing up for scheduled appointments, or receiving a service and asking if they can pay later at the end of the service. These are common scenarios that booth renters face, and that is why it is important to keep client partnerships professional and not personal.

Focus on the Client

In observing booth renters over a period of time and listening to their clients, it has been noticed that some stylists who booth rent are more focused on themselves than on their client. The independence of a booth renter who no longer has supervision or accountability can change the way customers are being serviced. This new-found freedom can appear to be more self-serving than client-focused (Figure 6–3). This self-serving attitude may manifest itself in such behaviors as showing up late for appointments, eating in front of clients, leaving clients in the chair while going out for a smoke, keeping clients in the salon all day, talking or texting on the cell phone while servicing customers, and leaving to take care of personal business while your clients wait. Remember, although you are your own boss, you are still running a business. Use your independence as a way to manage and serve clients at a high level of excellence and your reward will be greater in the end.

FYI
Many booth renters have been known to "shop hop" from one salon to the next approximately every six months to a year due to poor planning. When you move your business constantly and do not provide any stability with your business, it is hard to build client partnerships.

HERE'S A TIP
Send your customers a survey at the end of the year so they can evaluate your services as an independent beauty professional. Sending a survey will allow your clients to give you feedback on their service experience, and it will help you improve your services. An example of a survey is provided in the Quality Customer Service Survey (Exhibit A) at the end of the chapter.

Communicate with Clients

When creating client partnerships, stylist and client communication is a must. You must have clear communications with your clients to assure them that you understand what they want. In other words, your job is to find their need and fill it. For instance, a client may say to a hairdresser that she wants a trim. The stylist and the client's interpretation of a trim may differ. Without an effective consultation, the stylist may not give the client what he or she wants. This results in an unhappy

FIGURE 6–3

Focus on the Client, Not on Yourself

client who will not return due to miscommunication. In order to communicate effectively with your clients, make sure to:

- Always begin each service with a two- to three-minute consultation. If necessary, you may want to schedule a longer consultation with first-time clients. Doing so puts them at ease, creates a relaxed "getting to know you" environment, and ensures that you understand their wants and needs before providing the perfect service.
- Keep your portfolio, color swatches, and hair magazines handy to use during your consultation for greater visual accuracy of the client's wishes.
- Have your client sign a release form if you will be performing any chemical services. Refer to Exhibit B for a sample release.
- Gather your client's personal information, such as name, address, cell phone number, and e-mail address, and use technology such as e-mail, surveys, and social media to keep your clients current on new services and products (Figure 6–4).

Keeping Track of Your Clients

Since clients are the lifeline of your business, it is important that you gather information on every client you come in contact with. During their first service visit, have each client fill out a client information card. (NOTE: At the end of each day, if you use a salon software, be sure to enter all new client information or update any current client information.) When a client first calls to schedule an appointment, it is recommended you get the client's name, telephone number, and e-mail address. This is great information to gather by phone because if you are using a mobile application, the application can send out an e-mail or text to confirm the client's appointment. Once the client comes in for the appointment, they can then complete the client card. An emphasis is made here on the use of technology because as an independent beauty professional, you are the front and back of your operation, and the use of technology for tracking purposes makes your job much quicker and easier. If you do not have access to such technology, use Figure 6–5 as a guideline to the type of information to collect from your customers so as to successfully build a client partnership.

FIGURE 6–4

Use New Technology to Communicate with Clients

FIGURE 6-5 New Client Intake Form

CLIENT INTAKE FORM (SIDE 1)

Client's First Name: Last Name:

Address:

City: State: Zip Code:

Home Telephone: Cellular Phone:

E-mail Address:

Occupation:

How did you hear about us? ☐ Website ☐ Referral ☐ Flyer ☐ Billboard ☐ Other: _____

How often do you shampoo and condition your hair? ☐ Daily ☐ Weekly ☐ Bi-weekly ☐ Other: _____

How would you rate your hair's condition? ☐ Good ☐ Fair ☐ Poor ☐ Other: _____

Have you ever used haircolor? ☐ Yes ☐ No

When was your hair last trimmed?

Have you ever worn hair extensions? ☐ Yes ☐ No

If yes, please list what type? ☐ Sew in Weave ☐ Net Weave ☐ Bonding ☐ Hair Prosthesis
☐ Infusion ☐ Braids ☐ Other: _____

Is your scalp ☐ Oily ☐ Dry ☐ Flaky/Crusty ☐ Red/Inflamed ☐ Itchy
☐ None of the above

Please read below and check all that apply.
☐ I take hair nutritional vitamins. ☐ I had a hair transplant. ☐ I wear extensions.
☐ I am on prescription medications. ☐ I am under a physician's care. ☐ I am under stress.
☐ I eat a balanced diet. ☐ I drink plenty of water.

(Continued)

FIGURE 6–5 New Client Intake Form (*Continued*)

CLIENT INTAKE FORM (SIDE 2)		
Date of Service	Services Received	Notes & Formulas

Inform Clients of Products and Services

It is also important in the communication process to educate your clients on the products and services you offer. If there is a new service you are recommending to your customer, provide the client with as much information as you possibly can. If you have literature on the new service, provide your client with this literature during the consultation to help in your client's education process. Informing clients of new products and their ingredients will also help you during your retail product recommendation. Before you use a new product on a client, communicate to them what you are using, why you are using the product (product benefits), and how they should use the product at home. If a client understands the benefits of a product, the customer is more likely to purchase that product as a retail item at the end of the service. Retail will be discussed further in Chapter 7, "The Day-to-Day Details."

6.2 Establish Client Incentives

Offering client incentives is another form of communication to effectively create client partnerships. Client incentives make the client feel appreciated. When a client has a birthday or anniversary, you may choose to offer a discount on your services or products as a gift (Figure 6–6). When special holidays like Mother's Day, Father's Day, Valentine's Day, and so on come around, offer some kind of incentive to let your clients know you appreciate their business and that they are important to you. These small tokens of appreciation help with creating successful client partnerships.

FIGURE 6–6

Offer Special Discounts to Loyal Customers

6.3 Client Retention Strategies

If you keep your clients informed of what is going on in your business, you will build client loyalty, which is one client retention strategy. Other client retention strategies are as follows:

- Stay in constant communication with your clients through e-mails and social media outlets as mentioned in the marketing chapter. Ask all of your clients to join your Facebook (FB) fan page or to follow your business on Twitter. Request that they "check in" on their social sites upon arrival. Keeping your clients up to date on the latest hair cut, product, tools, and hair technology will keep your clients engaged and will help you retain their business (Figure 6–7).

- If you have not seen a client in a while, send them a card or pick up the phone and give them a call to see how they are doing. This small gesture shows the client that you appreciate them.

- Follow up with your customers after a new service is performed to find out how the service is working out for them, especially if the service was an expensive service like hair extensions.

- Offer loyalty cards to customers who come in on a consistent basis. The loyalty card should give the client a discount on his or her services as long as he or she receives a service from you a certain number of times a year.

FYI

To show you appreciate your clients and remember them, it is a good idea to purchase preprinted design postcards with a nice birthday or holiday message that you can send to them on special occasions. Cards that say "We Miss You" are great to send to customers you have not seen in a while. For more information on how to purchase these types of cards, visit salonstationerys.com.

FIGURE 6–7

Keep Clients Abreast on New Products

- Plan an activity for your clients. Host a client appreciation week and give a special discount or offer a free haircut with a color service during this time.
- During the holidays, invest in the people that invest in you. Pull your client reports, find the 10 clients who have booked the most number of services with you for the year, and give them a nice gift as a token of your appreciation. For your other clients, purchase a small inexpensive token of appreciation, gift wrap it and give it to these clients during the holidays. Clients always love gifts.
- Send out surveys to your customers to get feedback on how they feel about the services they receive and the environment of the salon.

Be creative and brainstorm about different ways to keep your clients engaged. The more a client stays active in your business, the longer you will retain them.

6.4 Coaching Corner: Business Ethics

In an employee-based salon, the salon manager handles any client issues. As a booth renter, you have to deal with all facets of the business, including customer complaints, sending out announcements and promotions, and performing the service.

At some point in your business, you will have to deal with an unhappy customer. When it comes to customer relations, the first thing you want to do if a customer becomes disgruntled is to try to resolve the conflict (Figure 6–8). If a customer becomes angry, do not become emotional as well. Instead, to effectively diffuse the situation, you should do the following:[1]

FIGURE 6–8

If a Customer Becomes Angry, the First Thing to Do Is to Resolve the Conflict

- Give the customer an opportunity to voice their concern.
- Pay attention to what the customer is saying.
- Repeat what the customer is saying to show them you understand.
- Express regret and sympathy for their frustration.
- Restore the customer's confidence by letting them know you are committed to resolving the problem.

- Stay positive.
- Stay professional.
- Tell clients specifically how you will resolve the issue.

Operating all the various aspects of your business can be a bit overwhelming, but one thing you must remember is that this is your business, and your goal is to be successful, so dealing with customer relations is just another part of the job that helps strengthen the client-partnership process.

6.5 Summary

Building and retaining clientele is a requirement for running a successful independent beauty business. Since this area is crucial to the success of your business, it is important your annual marketing plans are designed around clientele building and retention. Remember, a loyal client has the potential to bring more clients into your business; therefore, building client partnerships is important.

Let us review the top takeaways from "Strategies for Retention and Building Clientele."

6.6 Top Takeaways: Strategies for Retention and Building Clientele

- **Client partnerships.** Building client partnerships is an important part of growing your booth-rental business. The more client partnerships you build, the greater your potential for revenue growth.

- **Incentives.** During the client-partnership building process, it is important to understand that a client partnership is not a friendship, it is simply a relationship. Offering client incentives is a great way to communicate with your clients to keep them engaged in your business.

- **Follow-up.** Keeping track of your clients through the use of client cards or through the use of client tracking software is important when building client partnerships. Take note of the clients you have not seen in a while. Send them a card or pick up the phone to check in to see how they are doing. This small gesture shows the client you appreciate their business.

---

[1] Salon and Spa Management Tools. (2008). Clifton Park, NY: Milady, a part of Cengage Learning.

Chapter 6 Quiz: Strategies for Retention and Building Clientele

This chapter explores how to build and retain clientele. Answer the following questions to review what you have learned.

1. One method of building clientele that does not cost money is
   a. sending e-mails.
   b. giving out flyers.
   c. word of mouth.
   d. passing out business cards.

2. The definition of a client partnership is
   a. a client with whom you are friends.
   b. a client who you frequently service.
   c. a professional relationship built with a client.
   d. all of the above.

3. During the client-partnership development process, keep in mind that
   a. a client can become your friend.
   b. clients must have your personal home telephone number.
   c. client partnership is about focusing on the client.
   d. none of the above.

4. Creating a successful client partnership requires that you _____ with the client.
   a. become friends
   b. go out to dinner
   c. communicate
   d. all of the above

5. To gather client information from your customers, you should ask them to leave a business card.
   a. True
   b. False

6. What is the most effective tool to keep track of your client's information and appointments?
   a. Planner or calendar
   b. Appointment book
   c. Salon software
   d. Notebook

7. Offering client incentives is one way to keep your clients engaged in your business.
    a. True
    b. False

8. When servicing a new client for the first time, it is important that you do a _____.
    a. great shampoo service
    b. neck and shoulder massage
    c. consultation
    d. all of the above

9. List three ways you can effectively retain clients.
    1. _____
    2. _____
    3. _____

EXHIBIT A

QUALITY CUSTOMER SERVICE SURVEY

Date: _____

Client's Name: _____

Technician: _____

Rate the effectiveness of each item of action:	Effective	Average	Ineffective	N/A
1. General appeal of salon/spa (exterior and lobby)	❑	❑	❑	❑

2. Staff response
   a. Were you greeted *immediately*? ❑ Yes ❑ No
   b. By whom were you greeted initially? _____

3. Technician greeting
   a. Personal appearance:

	Effective	Average	Ineffective	N/A
▪ Professional clothes?	❑	❑	❑	❑
▪ Freshly styled hair?	❑	❑	❑	❑
▪ Cleanliness?	❑	❑	❑	❑

   b. When your technician met you, how effective was the technician's:

	Effective	Average	Ineffective	N/A
▪ Friendliness?	❑	❑	❑	❑
▪ Eye contact?	❑	❑	❑	❑
▪ Consultation?	❑	❑	❑	❑

4. Did you enjoy your experience? ❑ Yes ❑ No
   Why or why not?
   _____
   _____
   _____

5. What additional spa services would you like to see available to you?
   ❑ Massage   ❑ Facials   Other _____
   ❑ Body Treatments/Waxing ❑ Nail Services Other _____

6. What specialties would you like offered?
   ❑ Regular Coffee   ❑ Bottled Water   Other _____
   ❑ Flavored Coffee   ❑ Jazz Music   Other _____
   ❑ Juices   ❑ Classical Music   Other _____
   ❑ Flavored Teas   ❑ Easy Listening Music Other _____

7. When do you prefer to have your services? ❑ AM ❑ Midday ❑ PM ❑ Weekends

8. How could we serve you better?
   _____
   _____
   _____

chapter 7

The Day-to-Day Details

CHAPTER OUTLINE

7.1 Scheduling Appointments

7.2 Creating Services

7.3 Retailing

7.4 Pricing Your Retail

7.5 Coaching Corner: Formula for Success

7.6 Keeping Track of Your Product Inventory

7.7 Accepting Payments

7.8 Tracking Your Income

7.9 Summary

7.10 Top Takeaways: The Day-to-Day Details

Career Profile

Linda Garcia

Linda Garcia has been working independently on and off since 2001. Though booth renting gave her the ability to move around and start a family, it also made retaining her clientele and growing her business much more difficult. After years of trial and error, Linda has found herself in a successful position where she works independently, offering services at a local salon.

How did you become an independent stylist?

In 2001, I was happy working in a salon, but due to personal circumstances, the owner had to sell. So I started to consider my options. I was booked about a month in advance and had personal connections to many of my clients. By attending a hair show, I learned that a stylist needs at least 100 regular clients to be successful at booth renting. With 120 clients, I began the process by contacting each of them, ordering products, and gathering supplies.

What are some of the trials you experienced in booth renting?

Moving around is difficult when trying to remain independent. I moved 45 minutes away from the salon I was working at and lost about 25 percent of my clientele. I then moved to New Mexico and had to begin all over again. At first, I worked as a commission-based employee, building a clientele in my new area, and after four years, I decided to go back to booth renting.

How do you remain organized on a day-to-day basis?

My scheduling and bookkeeping stay organized through an online computer program called SpaBooker. I use a paper planner only for work. This planner has a place for receipts and also holds checks, cash, and other important documents. I have a journal where I write down important information, important numbers, business-related calls, and things I need. I do my best to write down client formulas, either during the processing time or right after they leave. SpaBooker and other computer programs like Salon Iris have a place to write formulas, but I prefer to store them on good old-fashioned index cards in a box. I do use a mobile device and salon software through SpaBooker. This program is crucial in staying organized and keeping track of inventory, if you sell retail. The program confirms appointments through e-mail and/or text. I make most of my appointments through my cell phone or texting. My clients also have the option to call the salon or use online booking through the salon website.

What are some strategies to maximize your business on a daily basis?

In the industry, as an independent stylist, you need to keep up to date with current services and styles. I suggest specializing in something you are excellent in, such as updos, color, or keratin treatments. You should research products, gather all the information you can, and be an expert at what you do. I also try not to overbook; if it is a new client, I would rather schedule more time for them so they do not feel rushed. For my regular clients, I know who takes more or less time and I book them accordingly. I do look ahead a few weeks to be sure my book is scheduled correctly to avoid overbooking. If there is a schedule conflict, I call and notify the client as soon as possible.

What is some general advice you would give to people considering booth renting?

My advice is to do research in your area on pricing and the services needed. Look at what your options are and what resources you have to help with startup costs. Are you successful at your current salon? If you decide this is for you, be a good booth renter by keeping your things neat and organized. You are an independent business owner, so you need to dress and act professionally. Continue your education with local classes and international hair shows. Use social media sites such as Facebook (FB) and Pinterest to stay connected with your clientele.

Working through the day-to-day details of your booth-rental business will come naturally to you as an independent beauty professional. Why? Because it is what you do every time you step on the salon floor to work. The difference between working as an employee in a salon and working as a booth renter is that as an employee, many facets of the day-to-day details are done for you by the salon. As a booth renter, you are responsible for working out every detail yourself. Working through the day-to-day operations of your booth-rental business alone requires a lot of organization and time-management skills. In this chapter, we will walk you through how to process the various daily details of your booth-rental business.

7.1 Scheduling Appointments

Scheduling appointments for customers is fairly straightforward. When a customer calls and wants to schedule an appointment, all you need to do is take down their personal information using a computer software program or printed appointment book, find out what services they want, confirm what day and time they would like to come, and then schedule the appointment, right? When it comes to scheduling appointments, the task is divided into two categories: accepting the call and scheduling the appointment.

Accepting the Call

Although scheduling an appointment is a simple process, the challenge for booth renters comes when customers are calling for appointments while you are servicing customers. Depending on the type of service you are performing, it can be challenging to take a call and schedule an appointment at the same time. When it comes to appointment scheduling, some booth renters answer their cell phone while servicing customers, while others let the phone ring to voicemail and call the client back later. It is not considered a safe or respectful practice to answer the phone and schedule appointments while servicing another customer, especially if you are performing a chemical service. One alternative to scheduling appointments as a booth renter is to allow your customers to schedule appointments through an online Web-based scheduling program (Figure 7–1). Online scheduling programs allow customers virtual access to your appointment book so they can book their own appointment. After the appointment is received online, you can call the customer back later to confirm the appointment.

FIGURE 7–1

Allow Clients to Book Their Appointments Online

Booking the Appointment

When you schedule an appointment to take a client, make sure you schedule enough time between customers to perform the service adequately. For instance, if a customer calls for an appointment to receive a color service at 10 am, allow enough time for this service to be performed, and for the customer to pick out any products, pay, and leave, before you schedule another service for the next customer. Often, booth renters forget they are a single operator and will overbook their schedule by booking clients 30 minutes apart. If you book clients 30 minutes apart and do not have an assistant, this will cause clients to have to sit and wait for long periods of time before you are able to service them. So make sure when you are scheduling appointments to take into consideration the time needed to perform the service before you schedule the next customer so that customers are not waiting.

Real Talk: Overbooking

One of the biggest complaints customers have about going to the salon is coming to the salon and waiting all day. Overbooking is a common problem amongst booth renters. If you overbook, you may get the money now, but you will not continue to get it much longer. Overbooking may happen if the time limit is incorrect for a service and then another customer is booked too soon, or if two or more customers are accidentally booked for the same time slot. If you want to keep your clients, and keep them coming back, try to think about what it would be like if *you* were the client and were forced to sit all day at the salon. It would not be a pleasant experience.

One key to being successful as an independent beauty professional is to understand that time is money and money is time. You can make more money by raising prices, increasing your clientele or increasing how often they visit you. If you have raised your prices, are consistently upselling, and are booked over 80 percent of your time, then you may be ready for an assistant. Hiring an assistant is one way to ensure that your clients get in and out of the salon in a timely manner. Think about when you go to the doctor or dentist for a visit. Before you see a doctor, you will always see the nurse first, and when you visit the dentist, you will always see the hygienist before the dentist. If you have a busy schedule, hire a shampoo assistant to help you with the servicing of your clients.

In order to prevent overbooking, review the following items:

- Look at your list of services, assign each service a length of time needed to fulfill the service, and schedule appointments accordingly.
- Hire an assistant to help with the flow of clients.
- If your budget does not allow for an assistant, when you schedule appointments leave an extra 30 to 60 minutes open before the next customer is booked.

Confirming Appointments

When it comes to confirming appointments, the best approach is to use a feature included in most salon software applications that is used to keep track of your clients via your computer or mobile device (Figure 7–2). This feature will automatically send an e-mail or text to the customer a day or two before the appointment to confirm their appointment date and time. If this use of technology is not available, you can call the customer by phone to confirm the appointment. To communicate with your customers regarding other special events, it is nice to send out e-mail blasts or postcards with special messages that offer discounts that are redeemable upon the presentation of that postcard or perhaps a printable coupon that you have sent via e-mail.

FIGURE 7–2

Client Communication with a Salon Software via Your Mobile Device Can Be Very Effective

7.2 Creating Services

In Chapter 3 in Exercise 3–7, titled "Booth-Renter Business Plan Template," you were provided a guide and template for how to create your service menu, along with how to input the prices of each service. Creating your service menu is relatively simple once you have learned how to price your services (Figure 7–3). In this section, you will place a time limit on each one of the services you offer. This is important because as you begin to schedule appointments with your customers, you will need to know how much time is needed to complete the service before another customer can be scheduled. Placing a time to perform a service on each of your services is also necessary if you decide to use a salon software or tablet application. The reason is these types of applications ask questions about service limits so that when an appointment is scheduled through the software, it will automatically block out the time before allowing you to schedule your next customer.

In Exercise 7–1, create your salon service menu and write down the time needed for you to complete each service. If necessary, use your pricing menu from Chapter 3, Exercise 3–7, "Booth-Renter Business Plan Template."

FIGURE 7–3

Create Your Service Menu Based on Your Expertise

EXERCISE 7–1 Create a Service Menu

SERVICE NAME	PRICE	TIME LIMIT
Example: Shampoo & Blow Dry	$50	30 minutes

7.3 Retailing

When it comes to retailing as a booth renter, there are only two things you need to concern yourself with: choosing the right line to retail and where you will store your inventory. Another important consideration is your plan for selling your retail. It will not fly off the shelf itself. If you are not effectively educating your clients on retail and recommending products for purchase, you will find that your retail simply collects dust. Dust does not pay the bills! Retail is an excellent opportunity to grow your revenue and it does not take any extra time.

Choosing the Right Line to Retail

With so many hair product lines available to you, how do you know what line to retail? The selection of the perfect line for you will depend on the main services you offer your customers (Figure 7–4). When choosing a retail line to sell, you must consider four things about the line:

1. **Brand exposure and personal brand alignment:** You want to first identify a retail line of products that has consumer exposure. It is easier to retail a line of products that has brand recognition to a customer. By brand recognition what is meant is a line of retail products designed for salons that are advertised in magazines, television commercials, and any other form of media that brings about consumer awareness. Additionally, you should consider whether or not the brand you choose is aligned with your personal business values and philosophies. You will have a much greater chance of selling your retail if you believe in it and are passionate about the value it brings to your customers.

2. **Meet the needs of the customer:** When considering a line of retail products to sell, think about the customers you service and their home maintenance needs. If the majority of your clients have color, you definitely want to find a line with high exposure that specializes in hair-color maintenance. If your primary service is makeup or nail services, you want to select colors and shades that fit the needs of your current clientele.

FIGURE 7–4

Selecting the Perfect Line to Retail Depends on the Type of Services You Perform

3. **Salon exclusivity:** Choosing a line of products your client cannot purchase from a retail store, but only from a salon, is common in the beauty industry. The decision of whether to sell salon-exclusive products or products in the general market is totally up to you. Salon-exclusive products seem to be the popular choice among most beauty industry professionals because they do not have to buy in large quantities to start retailing and it is easier to set up accounts with companies that sell products exclusively to salon professionals.
4. **Packaging:** This plays a major role in whether customers purchase retail or not. Customers love to purchase great products in nice packaging.

Where to Store Your Inventory

If your business plan requires retailing dollars to succeed, then it is important to find a salon to rent from that has stations with built-in retail shelving, or it must be communicated with the salon owner during the salon-selection interview process that your business model includes retail (Figure 7–5). During the research process, open up conversation with the owner to determine if a communal retail area is provided for the booth renters to store their retail.

Since retailing is also a visual process, it is important to find a salon that will allow you to show a visual of your products.

FIGURE 7–5

Planning on Retailing? Find a Salon to Rent from Whose Stations Have Built-in Retail Shelving

7.4 Pricing Your Retail

The pricing you create for your retail is very important to the bottom-line profit of your business (Figure 7–6). Pricing your retail is one of those areas that is improperly calculated most of the time. When pricing your retail, there are three things you must consider: the wholesale cost of goods, competitor pricing, and how much profit you are expecting. These three areas are very important to the formula in deciding how much you will charge for your retail. Let us explore these three areas.

Wholesale Cost of Goods

Wholesale cost of goods is the discounted price of an item purchased from a vendor to be sold for

FIGURE 7–6

Pricing Your Retail Correctly Is Important to the Bottom-Line Profit of Your Business

resale. Before you price your retail item, you must first take into account the cost you paid for the item you are intending to retail. When purchasing items for retail, make sure to ask for the wholesale/discounted cost of the items you are purchasing. Generally, before a wholesale cost is given to a business, you will be required to show some type of business documents such as a business license, a resale certificate, and/or a proof of professional license. Many stores that sell to beauty professionals only require a professional cosmetology, esthetician, or nail license to purchase items already priced for resale.

Competitor Pricing

As an independent stylist, it is best practice to research how much your competitors are charging for the same items you are planning to retail. Look to other salons in the area, as well as online websites that sell the same item, and then record three price levels for the item: the lowest price, the medium price, and the highest price. Analyzing three pricing levels for the item you are planning to retail will allow you to decide on which end of the pricing scale you want to be. Always remember, whatever side of the scale you choose for your retail pricing, make sure to consider the purchase price of the item first. Regardless, it is not the price that will sell your retail, it is your education and personal recommendation that builds your customer's desire to purchase it. Customers buy on the basis of good feelings and solutions, not price.

Expected Profit

Once you determine your purchase prices of items and look at your competitors' pricing for those same items, it is now time for you to set the retail pricing for your items. Before you do so, first determine how much profit you want to make on your retail items (Figure 7–7). In the standard retail market, the average profit made on a retail item can range from 30 to 50 percent. To determine your markup, you will need to decide the amount of profit you want. If you want 50 percent profit, you will mark up your retail by 50 percent, and so on. Let us use some real numbers to make this clear.

The formula to use for retail profit is

Cost of goods/(1 − Expected profit percentage)

For example, let us determine the retail cost of a shampoo item that you purchased at a wholesale cost

FIGURE 7–7

Before You Price Your Retail, Determine How Much Profit You Want to Make

for $7.50. Using a simple formula, the retail price of your shampoo at a 50 percent profit is:

$$\$7.50/(1 - 50\%) = \$15.00$$

You can see how using the retail profit formula to price your retail makes a big difference to the profit of your bottom line. Let us view some other sample equations for retail profit markups.

- 50% Profit = $7.50/(1 − 50%) = $15.00
- 45% Profit = $7.50/(1 − 45%) = $13.64
- 40% Profit = $7.50/(1 − 40%) = $12.50
- 35% Profit = $7.50/(1 − 35%) = $11.54
- 30% Profit = $7.50/(1 − 30%) = $10.71
- 25% Profit = $7.50/(1 − 25%) = $10.00
- 20% Profit = $7.50/(1 − 20%) = $9.38

Pricing your retail correctly can give a big boost to the revenue you generate daily. If you follow these three simple rules of retail pricing, you will see your income flourish.

7.5 Coaching Corner: Formula for Success

Many booth renters struggle with the independent business model once they make that transition into independence. Another challenge that booth renters face is being confident to stand on the pricing structures they set for their business. Unfortunately, it is a common practice for booth renters to be persuaded by their customers to lower their prices because they do not want to lose them. Some first-time renters, when moving from a commission salon to rental, will lower their prices, thinking this will encourage their clients to follow them. Either approach is a recipe for trouble. The minute you start wavering and show a lack of confidence is the minute that customers begin to ask for better pricing, deeper discounts, and so on. Do not be alarmed by price adjustment requests from customers. It may happen. If it does, you should confidently stand by your service and your price and not back down. Otherwise, it can become a slippery slope of confusion, multiple prices, and frustration.

FIGURE 7–8

Sample Service Policy

(Front)

(Back)

As a booth renter, what you must do is clearly state your company's policies upfront to the customer (Figure 7–8). You should always have a price menu clearly displayed at your station and you may also choose to provide each customer with a new client packet that will include a pricing menu, your business card, a brochure that includes an overview of the services you offer, and a description of your policies, including how to schedule and cancel appointments, and information about payment policies and your client privacy policy.

7.6 Keeping Track of Your Product Inventory

Keeping accurate track of your inventory is crucial to the success of your booth-rental business. As you go about your daily operations, you are constantly using backbar and retail inventory. So it is imperative that when you start your booth-rental business, you keep an accurate account of your inventory from the beginning (Figure 7–9). When you operate as an independent beauty professional, it is easy to overlook small things like inventory. You do not want to be in the middle of servicing a customer and have to go out and purchase items needed to complete the service while your client waits for your return. This is another area where booth renters have been found lacking. Borrowing products from others in the salon is not professional and can cause tension between stylists. It is also unprofessional to have your client stop off at a local beauty store to pick up the products needed.

The best way to keep track of your inventory is to use salon software for a tablet or mobile device. The salon software tool is great because when you sell retail to your customers, the item is automatically taken out of inventory and your inventory totals are updated. Salon software also has a reporting feature that allows you to pull an inventory report so you can identify what items are needed to replenish your inventory so you will not be without the products and tools you need. Again, if you do not have access to inventory, you can use an inventory tracker form. A common rule of thumb for purchasing retail products is to maintain a budget of 50 percent. For example, if you sold $100 dollars last week in retail, you should buy back at least $50 in order to replenish your stock. The same theory holds

FIGURE 7–9

Inventory Tracking Is Crucial to the Success of Your Business true for backbar items. Your budget for backbar purchases is no more than 5 percent of service sales. For instance, if your service sales for the week were $1,500, then you have up to $75 to spend on backbar items in order to ensure you have products to use on clients for the following week. Use the inventory tracking form shown in Exercise 7–2 to manually track your inventory.

EXERCISE 7–2   Inventory Tracking Form

ENTER SALON/SPA NAME

Step 1) ENTERING THE INFORMATION:

MONTH OF ANALYSIS

Step 2) ENTERING SALES AND PURCHASE INFORMATION:

Retail Percentages—Enter all "prior-week" retail sales and "this-week" retail purchases

	WEEK 1	WEEK 2	WEEK 3	WEEK 4	WEEK 5
Prior-Week Retail Sales:					
This-Week Retail Purchases:					
Budget 50%: Purchases divided by sales					

Backbar Percentages—Enter all "prior-week" Service Sales and "this-week" Backbar Purchases

	WEEK 1	WEEK 2	WEEK 3	WEEK 4	WEEK 5
Prior-Week Service Sales:					
This-Week Backbar Purchases:					
Budget 5%: Purchases divided by sales					

Step 3) RETAIL PROJECTIONS
Retail Budgeting—Avg. retail purchase GOAL is 50%

	MONTHLY	BUDGET	UNDER/OVER
Total Monthly Retail Sales:		50%	
Total Monthly Retail Purchases:		$ -	
Retail %: Purchase divided by sales			

Step 4) BACKBAR PROJECTIONS
Backbar Budgeting—Avg. backbar GOAL is 5%

	QUARTERLY	BUDGET	UNDER/OVER
Total Monthly Service Sales			
Total Monthly Backbar Purchases			
Backbar %: Purchases divided by sales			

7.7 Accepting Payments

When customers pay for their services, the most common payment methods are by credit card, cash, and checks. Many booth renters accept cash and checks for their services, but some still do not accept credit cards. According to Dave Ramsey's *The Money Answer Book*, studies have shown that clients will spend 12 to 18 percent more money and request extra salon services if they can use their credit or debit card to pay for them[1] (Figure 7–10). What this means is that as an independent beauty professional, you can increase your income and service sales by up to 18 percent if you accept credit cards as a

FIGURE 7–10

Accepting Credit Cards Can Increase Your Revenue by 12–18%

FYI

Companies like "The Square" and "Paypal" offer free credit-card swipers for your mobile device when you sign up to accept credit cards through their services. These companies charge a flat fee per transaction with no activation charge and no monthly service or statement fees. For more information about these credit-card service providers, visit squareup.com and paypal.com.

form of payment. The increased revenue opportunity far outweighs the small fees associated with the use of credit-card machines. Plus, customers find it inconvenient to have to visit an ATM machine before arriving at the salon for their services. If they can use a credit card to pay for their services, they are more likely to add on other services and purchase retail products from their stylist, especially if they do not have to worry about having enough cash to pay for the services.

Years ago, it was hard for booth renters to accept credit cards because of having to add a landline for the credit-card machine. Well, those days are long gone. Booth renters can now run their businesses more efficiently and increase their income and service sales by accepting credit cards in several different ways. Booth renters can buy a wireless credit-card terminal, or use their cell phone and purchase mobile applications or card-swipe tools like the iPhone card reader (at www.squareup.com.) that let them accept credit cards at a low cost.

Accepting credit cards is also a great tool to keep track of your daily and monthly revenue. When you accept credit cards, at the end of your business day your credit-card terminal will close out and give you a total of your sales for that day. You will also receive a monthly statement from your credit-card processor with a total of your monthly sales. If you want to increase your revenue and track your sales efficiently, accepting credit cards is a must.

7.8 Tracking Your Income

Keeping track of your income is an essential part of running the day-to-day operations of your business. One of the best ways to track your income is to use salon software that you can download onto your laptop, tablet, or mobile device. With today's world of technology, for about $25 a month you can run your entire business operation from your mobile device (Figure 7–11). Mobile salon software (after the initial setup of prices, services, and retail inventory) will keep track of your revenue daily to help you meet your income goal. Salon software will also allow you to pull reports to track your service and retail income daily, monthly, quarterly, and yearly. The reports made available through the program will also help you with your bookkeeping efforts because these reports can be handed directly to your bookkeeper for income-recording purposes.

Using salon software via your mobile device or tablet is one of the most efficient ways for you, the booth renter, to keep track of your income without spending a lot of time at the end of the day using calculators, pen, and paper to close out your day. Remember, you perform a one-man show, so use technology as your friend to help lighten the load of your day-to-day operations.

If you do not have access to much technology, you can purchase a recordkeeping appointment book, which lets you schedule appointments for your clients, as well as track the payments and tips you receive from each. This type of appointment book is an income journal designed to help you maintain an accurate register of your daily, weekly, and monthly service sales, tips, and retail sales. Otherwise, to keep track of your service and retail revenue daily, use the standard daily tracking form shown in Exercise 7–3. Use Exercise 7–4 to track your daily retail sales if you do not have access to technology.

FIGURE 7–11

Keep Track of Your Income with Software on a Mobile Device

HERE'S A TIP

If you are looking for a good mobile application that will work with your mobile tablet or smartphone and is designed with the booth renter in mind, the Vagaro Application is one of many choices available to you. This application lets you schedule appointments for customers as well as track your income. For more information, visit http://www.vagaro.com.

EXERCISE 7–3 A Service Income Tracking Form

Instructions: Use this form to track your daily income for services and tips. At the end of the day, add your total service costs and tip amounts and determine your total daily revenue.

SERVICE INCOME TRACKING FORM   DATE:_____

Customer Name	Service Received	Cost of Services	Tip Amount	Payment Type (Credit/Check/Cash)
1.		$	$	
2.		$	$	

(Continued)

EXERCISE 7–3 Service Income Tracking Form (*Continued*)

Customer Name	Service Received	Cost of Services	Tip Amount	Payment Type (Credit/Check/Cash)
3.		$	$	
4.		$	$	
5.		$	$	
6.		$	$	
7.		$	$	
8.		$	$	
9.		$	$	
10.		$	$	
11.		$	$	
12.		$	$	
13.		$	$	
TOTAL DAILY REVENUE		$	$	

Daily Summary: Total Cash $ _____ Total Checks $ _____ Total Credit Cards $ _____

EXERCISE 7–4 Weekly Retail Tracking Form

Instructions: Use this form to track your weekly retail revenue.

WEEKLY RETAIL TRACKING FORM   Week Ending _____

Date	Product Name	Amount	Method of Payment (Cash/Check/Credit Card)
		$	
		$	
		$	
		$	
		$	
		$	
		$	
		$	
		$	
		$	
		$	
		$	

(Continued)

EXERCISE 7-4	Weekly Retail Tracking Form (*Continued*)		
Date	Product Name	Amount	Method of Payment (Cash/Check/Credit Card)
		$	
		$	
		$	
Weekly Retail Sales Total		$	

RESOURCES

For digital spreadsheets to track your inventory purchases, weekly sales, and much more, check out the Financial Analysis and Coaching Tools page at http://www.milady.cengage.com. Once there, go to Salon/Spa Business Tools and search for *Financial Analysis and Coaching Tools*.

Although you can use the two preceding forms to track your service, tips, and retail revenue, it is highly recommended you streamline your operations and use technology as your virtual employee. This will make life simpler for you.

7.9 Summary

Now that we have covered the day-to-day details of the booth-rental operation, let us summarize them by reviewing our Top Takeaways for this chapter. Make sure all of these areas are covered at the end of your business day.

7.10 Top Takeaways: The Day-to-Day Details

- **Appointments.** Scheduling and booking appointments is an important part of your independent beauty business. When scheduling appointments, make sure to allow enough time between clients for the services requested so as not to keep your next clients from having to wait too long before they are serviced.

- **Retail.** When looking at items to retail, choose a retail line that has brand exposure and meets the needs of your clients, and that also matches your values and preferences.

- **Pricing retail.** To determine the correct pricing for retail items to be sold, consider your cost for the item, the suggested retail price for the item, and your expected profit.

- **Inventory tracking.** Keep an accurate account of how much retail and backbar product you have by tracking product inventory. Inventory tracking can be done manually or electronically. Whichever method you decide is up to you, just be sure to maintain your budgets and make certain you have enough product on hand for your day-to-day operations.

- **Credit-card payments.** Accepting credit cards can increase your revenue by at least 12 percent. There are several credit-card services that can be used in conjunction with your cell phone, where you do not have to pay monthly fees or purchase expensive equipment.

- **Track income.** It is important at the end of your business day to close out by totaling the service and retail income you have generated that day. Daily tracking and recording of your income is essential and will help simplify your bookkeeping.

---

[1] Ramsey, D. (2010). The Money Answer Book: Quick Answers for Your Everyday Financial Questions. Nashville, TN: Thomas Nelson.

Chapter 7 Quiz: The Day-to-Day Details

This chapter walks you through the day-to-day details of your booth-rental business. Answer the following questions to review what you have learned.

1. If your clients are calling your cell phone to schedule an appointment while you are servicing other clients, you should
    a. answer the phone and schedule the appointment.
    b. answer the phone and tell them you will call them right back.
    c. let the phone go to voicemail.
    d. none of the above.

2. Select one of the best ways for your clients to schedule appointments.
    a. Call the salon
    b. Call your cell phone
    c. Schedule an appointment online
    d. All of the above

3. If you have a busy schedule, the best way to help get clients out of the salon in a timely manner is to _____.
    a. cancel the client's appointment
    b. use an assistant
    c. reschedule the appointment
    d. None of the above

4. When pricing your retail, what three things must you consider:
    1. _____
    2. _____
    3. _____

5. A client may increase their service ticket if they can pay for their bill with a credit card.
    a. True
    b. False

6. Four factors in choosing the right retail line are
    1. _____
    2. _____
    3. _____
    4. _____

7. The best mechanism for tracking your clients, income, and inventory is to use (a)_____.
   a. recordkeeping appointment book
   b. tracking form
   c. salon software on a mobile device
   d. all of the above

8. Which of the following will NOT help you continually make more money?
   a. Raising prices
   b. Increasing your number of clients
   c. Booking two or more customers for the same time slot
   d. Increasing how often clients visit you

EXHIBIT A   The Day-to-Day Checklist

- ❏ Make sure you have enough cash to make change for cash payments.
- ❏ Schedule client appointments.
- ❏ Confirm client appointments for the next business day.
- ❏ Be sure your station is clean and disinfected, as well as the implements and supplies needed for the service.
- ❏ Check client in for service.
- ❏ Have client complete a client intake form if needed.
- ❏ Service clients in a timely manner.
- ❏ Upsell additional services and retail items.
- ❏ Record client services and input any custom formulations using salon software or client intake forms.
- ❏ Process client payments.
- ❏ Reschedule client for next visit.
- ❏ Check out client for the day.
- ❏ Clean your station for the next client.
- ❏ Record client payments and tips in your income tracker.
- ❏ Record retail sales in your retail-income tracker or software.
- ❏ Close out your credit-card machine.
- ❏ Clean your station for the end of the day.
- ❏ Check back-bar and retail inventory.

notes

appendix a
Self-Assessment

The purpose of writing this book is to help you properly set up and manage your booth-rental business and also to help you avoid the costly mistakes I made as an independent operator. Many of us when going into business had no one to help us get our business together, thus we were self-taught and had to learn about business the hard way. I hope the information you gained from this book will elevate your level of thinking and assist you in making the changes needed to make your business more successful. One of my mottos is "If I had to go through an experience just to help someone else, then the experience, no matter how painful, was worth it."

Now that you have read the information in this book, let me give you a self-assessment test to see how you have been doing in your business.

On each line, **rate yourself from 1 to 10**, with 1 being the lowest and 10 being the highest. BE HONEST!

ASSESSMENT	RATING 1–10
Getting Your Business Started—Is your business properly structured, and do you have an Employer Identification Number from the IRS?	
Cleanliness—Do you keep your station clean? Do you clean your station between each client?	
Service Sales—Do you upsell extra services to your clients when they come in for one service?	
Product Suggestion—At the end of a service, do you recommend a product for purchase to your client?	
Time Management—Do your clients have to wait long to be serviced when they come into the salon?	
Rebooking—At the end of a client's service, do you offer a schedule to your client for rebooking?	
Financial Management—Are you keeping an accurate track of your income using a mobile device or tracking log and also saving for retirement?	
Marketing—Do you have a marketing plan, and is it effective?	
Pricing Structure—Are you charging what you are worth?	
Continuing Education—Do you keep abreast of the latest products and happenings in the industry by attending classes?	
Client Consultation—Do you have good communication with your clients and understand what they desire for their hair, skin, makeup, and so on?	
TOTAL RATING	

<u>Score:</u>
- 90–100   Keep up the good work!
- 80–89   You are on the right track!
- 70–79   Some changes have to be made, but you are doing good.
- Less than 70 It's okay, because you purchased this book and I know you are going to make a change to your business.

Appendix B

Answers to End-of-Chapter Quizzes

CHAPTER 1 ANSWERS The Basics of Running a Booth-Rental Operation

1. You know that you are ready to operate a booth-rental business if you
   a. serve many clients.
   b. make over $500 a week in services and sales.
   c. have your business and personal finances in order.
   d. all of the above

2. Booth renting is the perfect business model for students just graduating from beauty school.
   a. True
   b. False

3. List five responsibilities that you would have as a booth renter.
   1. Keeping your license(s) up to date.
   2. Setting your own hours.
   3. Scheduling your own appointments.
   4. Providing your own phone service.
   5. Setting your own prices.

4. After a client has received your services, the client should pay the _____.
   a. front-desk receptionist
   b. independent stylist
   c. salon owner
   d. none of the above

5. The person responsible for keeping your independent booth station clean is the _____.
   a. salon owner
   b. salon assistant
   c. independent stylist
   d. all of the above

6. The type of insurance that pays a claim when a client is injured while a booth renter is performing services for that client is _____.
   a. professional liability insurance
   b. health insurance
   c. general liability insurance
   d. occupational safety and medical insurance

7. An independent contractor should have _____ insurance, which will pay a claim if the independent contractor becomes disabled and cannot physically work for a period of time.
   a. dental
   b. disability
   c. health
   d. workers' compensation

8. The person responsible for paying employment taxes for a booth renter is the salon owner.
   a. True
   b. False

9. The salon owner is responsible for marketing and helping a booth renter build his or her clientele.
   a. True
   b. False

10. General liability insurance will protect a booth renter from any lawsuits, such as one that arises if a client has a serious allergic reaction from a chemical process.
    a. True
    b. False

11. Professional liability insurance will protect a booth renter in the event that someone is injured at the booth renter's place of business.
    a. True
    b. False

CHAPTER 2 ANSWERS Getting Your Business Off the Ground and Running

1. Which one of the following credentials is necessary for independent stylists to practice cosmetology in the United States?
   a. Federal tax ID number
   b. Cosmetology license
   c. Business license
   d. Sales tax ID number
   e. All of the above

2. The purpose for obtaining a sales tax ID number is to_____.
   a. buy salon supplies
   b. sell retail products
   c. cover equipment costs
   d. none of the above

3. A business license is required to operate an independent beauty business.
   a. True
   b. False

4. Starting a booth-rental business requires that you choose a business structure.
   a. True
   b. False

5. List the four primary business structures used in for-profit businesses.
   1. **Sole proprietor**
   2. **Limited liability corporation**
   3. **S corporation**
   4. **C corporation**

6. If you do not choose a business structure when starting your business, when taxes are due, your business will be classified as this business structure type:
   Sole proprietor.

7. Which business structure pays taxes twice?
   a. Sole proprietor
   b. C corporation
   c. S corporation
   d. Limited liability corporation

8. When searching for a salon in which to rent a space, there are five things to consider. List these in the space below.
   1. **Existing clientele**
   2. **Location**
   3. **Salon environment**
   4. **Salon floor plan**
   5. **Salon amenities**

9. A salon suite is an enclosed floor plan in which an independent beauty professional is assigned his or her own private space with a lockable entrance.
   a. True
   b. False

10. The relationship between a booth renter and the salon owner is a tenant–landlord relationship.
    a. True
    b. False

11. An _____ is defined as someone whose work is controlled by the terms of employment.
    a. independent contractor
    b. employee

CHAPTER 3 ANSWERS Eliminating Detours and Distractions with a Solid Business Model

1. The purpose of a mission statement is to _____.
   a. talk about your business structure
   b. discuss your pricing structure
   c. explain why you decided to open your business
   d. all of the above

2. A vision statement _____.
   a. is a short-term picture of your business
   b. states the concerns you have for your company in the years to come
   c. specifies where you want your company to be in the years to come
   d. all of the above

3. List two types of goals you should have for your business.
   1. **Short-term goals**
   2. **Long-term goals**

4. Your short-term goals should be S.M.A.R.T. Spell out the acronym for S.M.A.R.T.
   <u>Specific</u>
   <u>Measurable</u>
   <u>Achievable</u>
   <u>Realistic</u>
   <u>Time-specific</u>

5. Your long-term goals should include _____.
   a. service
   b. community
   c. revenue
   d. expansion
   <u>e. all of the above</u>

6. A business plan and a business model are the same.
   a. True
   <u>b. False</u>

7. A business plan defines how your business will make money.
   a. True
   <u>b. False</u>

8. A _____ gives you specifics about your business.
   <u>a. business plan</u>
   b. business model

9. When creating your pricing structure, you should consider _____.
   a. location
   b. salon amenities
   c. experience
   <u>d. all of the above</u>

10. List five things that will help your business grow.
    <u>1. Accepting all forms of payment</u>
    <u>2. Retailing</u>
    <u>3. Charging what you are worth</u>
    <u>4. Creating a successful marketing plan</u>
    <u>5. Hiring an assistant</u>

11. List four components of a S.W.O.T. analysis.
    <u>1. Strengths</u>
    <u>2. Weaknesses</u>
    <u>3. Opportunities</u>
    <u>4. Threats</u>

CHAPTER 4 ANSWERS Your Money, Your Future

1. List at least five categories that you can use for tax write-offs.
   <u>1. Advertising</u>
   <u>2. Mileage</u>
   <u>3. Insurance</u>
   <u>4. Rent and lease</u>
   <u>5. Office expense</u>
   (For more answers, see Chapter 4, Section 4.8, "Taxes")

2. List five things your business success will depend on.
   <u>1. Cash flow</u>
   <u>2. Building a clientele</u>
   <u>3. Retaining clients</u>
   <u>4. Scheduling and available work hours</u>
   <u>5. Personal expenses</u>

3. List three members you should have on your financial dream team.
   <u>Answers may include:</u>
   <u>a. Bookkeeper</u>
   <u>b. Accountant</u>
   <u>c. Financial planner</u>
   <u>d. Personal banker</u>
   <u>e. Insurance agent</u>

4. You can write off the miles driven to and from any hair trade show or beauty-supply store.
   <u>1. True</u>
   2. False

5. What is the short name of the retirement fund that you can set up for yourself as an independent contractor?
   <u>a. SEP account</u>
   b. SET account
   c. SIP account
   d. None of the above

6. What tax category would you use to write off fees paid to renew your cosmetology certificates?
   a. Advertising
   b. Insurance
   <u>c. Taxes and licenses</u>
   d. All of the above

7. As an independent contractor, which of the following are supplies you can write off for tax purposes?
   a. Shampoo, conditioners, and haircolor
   b. Blow dryers and curling irons
   c. Combs and brushes
   d. **All of the above**

8. You must keep receipts for everything you write off for taxes, except your cell phone bill.
   a. True
   b. **False**

CHAPTER 5 ANSWERS Marketing Your Booth-Rental Business

1. List three objectives of a good brand.
   Answers may include:
   a. Conveys the message of the brand clearly
   b. Validates the brands credibility
   c. Draws the prospective buyer emotionally
   d. Inspires the consumer
   e. Builds brand loyalty

2. A marketing plan and a marketing strategy are the same.
   a. True
   b. **False**

3. Which of the following item(s) will explain your advertising and marketing goals for a calendar year?
   a. A business plan
   b. A marketing strategy
   c. **A marketing plan**
   d. None of the above

4. A good marketing plan will include (a) _____.
   a. detailed budget
   b. marketing goals
   c. marketing strategy
   d. **all of the above**

5. List four cost-effective ways to market your business.
   1. Website
   2. Direct mail
   3. Social media outlets
   4. E-mail blasts

6. You can use _____ to list the prices and services you offer, display photos of your hair styles, and book appointments.
   a. the newspaper
   b. **your website**
   c. radio and television
   d. all of the above

7. Participating in photo shoots will not help you _____.
   a. **completely eliminate your competition**
   b. build your clientele
   c. advertise your different services and specialty hair designs
   d. build your portfolio to send to recording studios, television networks, and hair product companies

8. Which technique is becoming one of the more popular ways to market your business?
   a. Direct mail campaigns
   b. Billboard advertising
   c. **Text and mobile marketing**
   d. Television and radio advertising

9. Which social media icons should you include on your website and print materials?
   a. Facebook
   b. Twitter
   c. YouTube
   d. **All of the above**

10. When creating a promotion for a product or service, what do you do first?
    a. Select the promotion offering and then the marketing method.
    b. **Select the marketing method and then the promotional offering.**

CHAPTER 6 ANSWERS Strategies for Retention and Building Clientele

1. One method of building clientele that does not cost money is
   a. sending e-mails.
   b. giving out flyers.
   c. **word of mouth.**
   d. passing out business cards.

2. The definition of a client partnership is
   a. a client with whom you are friends.
   b. a client whom you frequently service.
   c. <u>a professional relationship built with a client.</u>
   d. all of the above

3. During the client-partnership development process, keep in mind that
   a. a client can become your friend.
   b. clients must have your personal home telephone number.
   c. <u>client partnership is about focusing on the client.</u>
   d. none of the above

4. Creating a successful client partnership requires that you _____ with the client.
   a. become friends
   b. go out to dinner
   c. <u>communicate</u>
   d. all of the above

5. To gather client information from your customers, you should ask them to leave a business card.
   a. True
   b. <u>False</u>

6. What is the most effective tool to keep track of your client's information and appointments?
   a. Planner or calendar
   b. Appointment book
   c. <u>Salon software</u>
   d. Notebook

7. Offering client incentives is one way to keep your clients engaged in your business.
   a. <u>True</u>
   b. False

8. When servicing a new client for the first time, it is important that you do a _____.
   a. great shampoo service
   b. neck and shoulder massage
   c. <u>consultation</u>
   d. all of the above

9. List three ways you can effectively retain clients.
   <u>Answers may include:</u>
   a. <u>Stay in constant communication through e-mails and social media outlets.</u>
   b. <u>Send them a card or call to see how they are doing.</u>
   c. <u>Offer client incentives.</u>
   d. <u>Offer a client loyalty card.</u>
   e. <u>Sponsor a client appreciation week.</u>
   f. <u>Follow up with your customers after a new service is performed to find out how the service is working out for them.</u>
   g. <u>Send out surveys to your customers to get feedback on how they feel about the services they receive and the environment of the salon.</u>

CHAPTER 7 ANSWERS The Day-to-Day Details

1. If your clients are calling your cell phone to schedule an appointment while you are servicing other clients, you should
   a. answer the phone and schedule the appointment.
   b. answer the phone and tell them you will call them right back.
   c. <u>let the phone go to voicemail.</u>
   d. none of the above

2. Select one of the best ways for your clients to schedule appointments.
   a. Call the salon.
   b. Call your cell phone.
   c. <u>Schedule an appointment online.</u>
   d. All of the above

3. If you have a busy schedule, the best way to help get clients out of the salon in a timely manner is to _____.
   a. cancel the client's appointment
   b. <u>use an assistant</u>
   c. reschedule the appointment
   d. none of the above

4. When pricing your retail, what three things must you consider:
   1. <u>Wholesale cost of goods</u>
   2. <u>Competitor pricing</u>
   3. <u>Expected profit</u>

5. A client may increase their service ticket if they can pay for their bill with a credit card.
   a. <u>True</u>
   b. False

6. Four factors in choosing the right retail line are
   1. <u>Brand exposure</u>
   2. <u>Meets the needs of the customers</u>
   3. <u>Salon exclusivity</u>
   4. <u>Packaging</u>

7. The best mechanism for tracking your clients, income, and inventory is (a)_____.
   a. recordkeeping appointment book
   b. tracking form
   c. <u>salon software on a mobile device</u>
   d. all of the above

8. Which of the following will NOT help you continually make more money?
   a. Raising prices
   b. Increasing your number of clients
   c. <u>Booking two or more customers for the same time slot</u>
   d. Increasing how often clients visit you

glossary

accountant a person who audits and inspects the financial records of a business and prepares financial and tax reports.[1]

Articles of Incorporation a legal document filed with the state government that confirms the creation of a corporation. This document includes the company address, and stock and shareholder information.

Articles of Organization a document required by the state government to file an LLC.

asset anything of worth on a balance sheet that is owned by or payable to the company.

balance sheet a picture of your company's financial situation over a certain period of time.[2] It shows the company's assets, liabilities, net worth, and equity.

bookkeeper a person skilled in the area of accounting who keeps track and records the financial transactions of a business.

booth renter also referred to as *independent contractor*; defined by the Internal Revenue Service (IRS) as "someone who leases space from an existing business and operates their own business as an independent contractor. A booth renter or independent contractor is responsible for his/her own record-keeping and timely filing of returns and payments of taxes related to their business."[3]

brand is a unique name, logo, mark, phrase, or a combination of these items that businesses use to differentiate themselves from other businesses in the market.[4]

business license a permit issued by the state government that allows individuals or companies to conduct business within the government's geographical jurisdiction.

business model defines how your business will make money by focusing on the products and services that generate the most revenue, and shows where your business will make the greatest profit.

business plan a written description of your business as you see it today, and as you see it in the next five years (detailed by year).

cash flow the amount of money coming in as revenue and going out of your business for expenses.

C corporation a corporation that is taxed separately from its owners. A C corporation's earnings are taxed twice.

client partnerships professional relationships built with a client to ensure customer satisfaction. In simple terms, it means that you will work to develop and nurture the relationship with your customer by communicating and providing incentives that will keep your customers engaged in your operations.

competitive analysis points out your competitors and takes a look at their business tactics to identify their strength and weaknesses as compared to your business.[5]

core business values a set of principles and beliefs that your company is founded upon and which represent what your company believes.

current assets assets owned by the company that are expected to be converted into cash in less than one year.[6]

disability income insurance insurance that pays a percentage of your income in the event you become disabled or unable to work due to injury, illness, or any other cause that will not let you perform your normal work activities.

employee a person whose work is controlled by the organization of employment.

equity the amount of money contributed by the owners and the amount of ownership one has in the company after all its debts have been paid off.

estimated tax the method used to pay tax on income that is not subject to withholding,[7] such as income you receive from customers for services rendered or earnings you receive from outside sources for services rendered as an independent contractor.

financial planner a skilled consultant who assists individuals and businesses in achieving their long-term financial goals by analyzing the client's current financials and designing a plan to help the client achieve those goals.[8]

fixed assets monies used to buy physical assets which are employed to produce income and which will not be turned into cash within a year's time.

income statement also known as a *profit and loss (P&L) statement*; explains all the income and expenses generated by the business over a period of time and depicts how profitable a company is.

independent contractor also referred to as a *booth renter*; a self-employed individual who provides a service for an organization, but whom the organization has no control over regarding what type of services they perform and how they are performed.[9]

insurance agent a representative of a state board–licensed insurance agency that sells insurance policies to customers.

intangible assets assets of an undetermined life expectancy that cannot be sold, even though they are neither material nor physical.

liabilities monies on a balance sheet that the company owes to creditors, such as loans, credit-card debt, and mortgages.

limited liability company (LLC) defined by the IRS as "a business structure allowed by state statute."[10] Allows for single-member ownership of a company in many states, but which gives the owner the legal protection of a corporation.

logo a recognizable and distinctive graphic design, stylized name, unique symbol, or other device for identifying an organization.[11]

long-term goals goals you plan to achieve for your business within three to five years.

marketing plan explains your advertising and marketing goals for a calendar year.[12]

marketing strategy supplies the goals and tactics for your marketing plans.[13]

mission statement a statement that describes who you are and what you do. Your mission statement should briefly describe the "what," "why," and "how" of your business.

negative cash flow when you have more money going out of your business to pay expenses than you have coming in.

net worth the value of a company where the company's assets exceed the company's liabilities.[14]

open-floor-plan salon a salon that is designed in an open space where beauty professionals share a common work space, with a common shampoo, color, and dryer area.

personal banker an employee of a financial institution who helps clients manage their assets, such as savings, checking, and money-market accounts, and mortgages.

positive cash flow when you have more revenue coming into your business than money going out for expenses.

professional liability insurance (professional indemnity insurance) a specific type of insurance designed for independent contractors that provides protection in the event of a professional mistake or any injury a client sustains during a service.

registered agent a person who is appointed to receive and send legal documents on behalf of the business.

Sales Tax ID Number (Sales Tax Exemption Certificate) a certificate received from your local state government that allows you to charge and collect local sales taxes on products and goods sold to customers.

salon suites a collection of mini-salons all under one roof, similar to a doctor's office building.

S corporation defined by the IRS as "a corporation that elects to pass corporate income, losses, deductions, and credit through to their shareholders for federal tax purposes."[15]

self-employment tax social security and Medicare payroll taxes paid by an individual contractor.

short-term goals goals you set for your business within the first two years of its launch.

sole proprietor the sole owner of a business who pays personal income tax on the earnings received from that business.[16]

S.W.O.T. Analysis a planning method used to identify the strengths, weaknesses, opportunities, and threats of your business.[17]

vision statement a statement that describes your company's goals, values, and direction. Your vision statement should state the dream you have for your business in the years to come and what it will look like when it gets there.

wholesale cost of goods the discounted price of an item purchased from a vendor to be sold for resale.

---

[1] Accountant. (2000). *The American Heritage® dictionary of the English language*, fourth edition. Houghton Mifflin Company. Retrieved from http://www.thefreedictionary.com/accountant.

[2] http://www.sba.gov/content/financial-statements#Balance Sheet.

[3] U.S. Department of Treasury, IRS. (2011). *Tax tips for the cosmetology & barber industry* (4902) [Brochure]. Washington, DC: U.S. Government Printing Office, p. 6. Retrieved from http://www.irs.gov/.

[4] http://www.investopedia.com/terms/b/brand.asp#axzz281AEc4zm.

[5] Definition of competitive analysis [online]. (2012). *Entrepreneur Media, Inc.* Retrieved from http://www.entrepreneur.com/encyclopedia/term/82078.html.

[6] http://www.investopedia.com/terms/c/currentassets.asp#axzz26eSvSn1u.

[7] IRS Publication 4902. *Docstoc.com*.Documents, Templates ... (n.d.). Retrieved from http://www.docstoc.com/docs/74518597/IRS-Publication-4902.

[8] Financial planner definition. (n.d.). *Investopedia* [online]. Retrieved from http://www.investopedia.com/terms/f/financialplanner.asp#axzz26eSvSn1u.

[9] IRS. (2012). *Independent contractor defined*. Retrieved from http://www.irs.gov/businesses/small/article/0,,id=179115,00.html.

[10] U.S. Department of Treasury, IRS (2012). Independent contractor defined. Retrieved from http://www.irs.gov/.

[11] Logo. (n.d.). BusinessDictionary.com. Retrieved from http://www.businessdictionary.com/definition/logo.html.

[12] http://www.entrepreneur.com/encyclopedia/term/82450.html.

[13] http://www.businessdictionary.com/definition/marketing-strategy.html.

[14] http://www.investopedia.com/terms/n/networth.asp#axzz26eSvSn1u.

[15] U.S. Department of Treasury, IRS (2012.). *S-Corporations*. Retrieved from http://www.irs.gov.

[16] Definition of Sole Proprietorship. *Investopedia.com*. Retrieved from http://www.investopedia.com/terms/s/soleproprietorship.asp#axzz2AexUFb3w.

[17] Mind Tools Ltd. (n.d.). *SWOT Analysis: Discover new opportunities, manage and eliminate threats*. Retrieved from http://www.mindtools.com/pages/article/newTMC_05.htm.

index

Accountant, 102
Account payables, 105
Account receivables, 107
Add-on services, 134
Advertising
   as tax-deductible category, 110
   tracking form, 140
Appointment(s)
   booking, 157–158
   confirming, 159
   overbooking, 158
   scheduling, 12, 157–159
Articles of incorporation, 34, 35, 104
Articles of organization, 33
Assets
   current, 107
   defined, 107
   fixed, 108
   intangible, 108
Assistants, hiring, 20–21, 22, 87

Balance sheet, 102, 105, 107
Bank deposit totals, 105
Bank statement reconciliation, 105
Base price worksheet, 82–84
Billboards, 132–133
BizOn Track, 69
Booking appointments, 157–158. *See also* Overbooking appointments
Bookkeeper, 102
Bookkeeping, 18
   reconciliation, 105
Booth-rental payments, 105
Booth renter
   business plan template, 76–80
   defined, 3
   legal rights as, 47–48
   personal income and expenses worksheet, 8–10
   readiness worksheet, 6–7
   responsibilities of, 11–21, 25
Booth renting
   as career choice, 4–5
   ideal location for, finding, 39–46
   insurance coverage, 21
   transition to, 21

Brand
   defined, 121
   goals of, 121
   understanding, 121–122
Budget, 123
Business cards, 19, 20
Business credit, building, 113–115
Business Credit USA, 115
Business ethics, 150–151
Business-expense tracker, 26–27
Business license, 39
Business model, 61–94
   components of, 63–64
   creation of, 74–80
   defined, 63
   goals of, 68–70, 80
   growth of, 80–87
   mission statement, 67, 73
   S.W.O.T. analysis, 88–91
   template, 64–66, 76–80
   value identification, 70–73
   vision statement, 67, 73
Business plan
   defined, 63
   executive summary, 74
   template, 76–80
Business structure, choosing, 31–37
   C corporation, 34–45
   corporation structures, examples of, 35–36
   determination chart, 37
   limited liability corporation, 33
   S corporation, 34
   sole proprietorship, 31–33

Call, accepting, 157
Cash, 107
Cash flow, 97–98
   defined, 97
   negative, 97
   positive, 97
Cash on hand, 104
C corporation, 34–45, 114
   pros and cons of, 35
Checklist
   day-to-day, 176
   financial health, 104–106

Clientele, existing, 40
Client incentives, establishing, 149
Client partnerships, 143–148
   communication with clients, 145–146
   distinguished from friendship, 144–145
   focus on clients, 145
   keeping track of clients, 146
   products and services to clients, informing, 148
Client retention strategies, 149–150
Coca-Cola, 121
Commercial space, for rent, 45–46
Communication, with clients, 145–146
Company description, 74
Competitive analysis, 74
Competitor pricing, 163
Confirming appointments, 159
Constant Contact, 129
Core business values
   defined, 70
   identification of, 70–73
   list, 71
   statement, 72
Corporation structures, examples of, 35–36
Coscioni, Edda, 62
Cosmetology license, 38
Costs, 97–99
Credit-card terminal batching, 105
Current assets, 107

Daily financial checklist, 104–105
Day-to-day checklist, 176
Day-to-day details, 155–176
   accepting payments, 167–168
   creating services, 159–160
   income, tracking, 168–172
   product inventory, tracking, 165–167
   retailing, 161–164
   retail pricing, 162–164
   scheduling appointments, 157–159
   success formula, 164–165
Determination chart, of business structure, 37
Direct mail marketing, 130–131
Disability income insurance, 16–17
Doing-business-as (DBA), 31
Dun & Bradstreet (D&B), 114, 115

E-mail marketing, 129
Employee, defined, 48
Employer identification number (EIN), 38–39
   application for, 52

End-of-year financial checklist, 106
Entertainment, as tax-deductible category, 112
Equifax Business, 115
Equity, 108
Estimated tax, 17
Executive summary, of business plan, 74
Expenses
   office, 111
   personal, 98–99
   prepaid, 107
Experian Business, 115
Experience, 81–82
Expertise, 81–82

Facebook (FB), 120, 123, 130, 133, 149, 156
Federal tax ID number. *See* Employer identification number (EIN)
Financial dream team, establishing, 101–103
   accountant, 102
   bookkeeper, 102
   financial planner, 102–103
   insurance agent, 103
   personal banker, 103
Financial health, analyzing, 106–108
Financial health checklist, creating
   daily checklist, 104–105
   end-of-year checklist, 106
   monthly checklist, 105
   quarterly checklist, 106
   weekly checklist, 105
Financial planner, 102–103
Financial projections, solid business model, 75
Fixed assets, 108
Frequent-buyer cards
   hand out, 134
   service, 134
Friendship versus client partnership, 144–145
Funding request, solid business model, 75

Garcia, Linda, 156
Goals
   of business model, 68–70, 80
   of good brand, 121
   income, establishing, 99, 101
Google Voice, 13

Hair Style, 132
Hand out retail frequent-buyer cards, 134
Hype Hair, 132

Ideal location for booth renting, finding, 39–46, 81
Income goals, establishing, 99, 101
Income statement, 102, 105, 107, 108
Income, tracking, 168–172
Independent contractor. *See* Booth renter
Individual health benefit plan, 108–109
Infection control, 19
Insurance
   coverage, 21
   disability income, 16–17
   professional liability, 15–16
   as tax-deductible category, 110
Insurance agent, 103
Insuring Style, 130
Intangible assets, 108
Interest, as tax-deductible category, 110–111
Internal Revenue Service (IRS), 3, 15, 17, 34, 102, 106, 109
Intuit, 103
Inventory, 105, 107
   store, 162
   tracking, 165–167

Keeping track, of clients, 146
Kentucky Fried Chicken (KFC), 121

Latest hair trends, keeping up, 86
Lease, as tax-deductible category, 111
LeClear, Sandy, 30
Liabilities, 108
License
   business, 39
   cosmetology, 38
   federal tax ID number, 38–39
   renewal, 11
   sales tax ID number, 39
   taxes and, 112
Limited liability company (LLC), 33, 36, 114
   pros and cons of, 33
Location, 81
Logo, 121
Long-term goals, 68–70, 80
Lovell, Sara, 2–3

Mail Chimp, 129
Maintenance, as tax-deductible category, 111
Management, of business model, 75
Market analysis, 74

Marketing, 19–20
   direct mail, 130–131
   e-mail, 129
   goals, 122–123
   mobile, 129
   and promotion strategies, 128–135
   text, 129
Marketing and promotions calendar template, 135–136
Marketing plan, 75, 122–128
   defined, 122
   successful, creating, 87
   template, 124–126
Marketing strategy, 122–128
   defined, 123
   evaluation form, 127
   evaluation of, 123–124
McDonald's, 121
Meals, as tax-deductible category, 112
Medina, Dawn, 120
Milady's Financial Analysis and Coaching Tools, 106
Mileage, as tax-deductible category, 110
Mission statement, 67, 73
Mobile marketing, 129
Monthly financial checklist, 105

Negative cash flow, 97
Net worth, 108
New client intake form, 147–148

Office expenses, as tax-deductible category, 111
Online portfolio, 128–129
Open-floor-plan salons, 44, 47
Operator tools, 15
Organization, of business model, 75
Overbooking appointments, 158. *See also* Booking appointments

Payments
   accepting, 167–168
   forms of, 85
Payroll recordings, 105
Personal banker, 103
Personal cash-flow evaluator, 100
Personal expenses, 82, 84–85, 98–99
   evaluation of, 99
Personal income and expenses worksheet, 8–10
Personal income goals, 82, 84–85

Phone service, 12–13
Photo shoots, 131–132
Pinterest, 156
Platform artist, 133
Positive attitude, keeping, 86
Positive cash flow, 97
Prepaid expenses, 107
Pricing
    base price worksheet, 82–84
    competitor, 163
    retail, 162–164
    for services, setting, 13–14
Print media, 131–132
Product(s)
    to clients, informing, 148
    inventory, tracking, 165–167
    solid business model, 75
    supplies, 14–15
Professional Beauty Association, 4, 130
Professional indemnity insurance.
    See Professional liability insurance
Professionalism, 86
Professional liability insurance, 15–16
Profit and loss (P&L) statement. *See* Income statement
Profit, retail, 163–164
Promotional offerings, 133–135
Promotion strategies, 128–135
    billboards, 132–133
    direct mail marketing, 130–131
    e-mail marketing, 129
    mobile marketing, 129
    offerings, 133–135
    online portfolio, 128–129
    photo shoots, 131–132
    platform artist, 133
    print media, 131–132
    radio commercials, 132
    social media outlets, 130
    technical educator, 133
    television commercials, 132
    text marketing, 129
    website, 128–129

Quality Customer Service Survey, 145, 154
Quarterly financial checklist, 106
QuickBooks, 103

Radio commercials, 132
Ramsey, Dave, 167

Readiness worksheet, 6–7
Rebooking, 134
Referrals, 135
Registered agent, 33
Rent, 18
    commercial space for, 45–46
    as tax-deductible category, 111
Repair, as tax-deductible category, 111
Retail(ing), 86
    pricing, 162–164
    profit, 163–164
    promotional ideas, 134
    revenue, 104
    right line to, choosing, 161–162
    sales tax report, 105
Retirement
    plan, 18
    saving for, 112–113
Revenue
    retail, 104
    service, 104
Rewards program, 135
Romney, Ramin, 96

Sales tax exemption certificate, 39
Sales tax ID number, 39
Salon(s)
    amenities, 41, 81
    environment, 41
    floor plan, 41
    open-floor-plan, 44, 47
    search comparison chart, 42–43
Salon suites, 44–45
    benefits of operating in, 45
    setup, 45
Saving, for retirement, 112–113
Scheduling appointments, 157–159
S corporation, 34, 36, 114
    pros and cons of, 34
Self-assessment, 178–179
Self-employment taxes, 17
Services
    to clients, informing, 148
    creating, 159–160
    frequent-buyer card, 134
    income tracking form, 169–170
    pricing for, setting, 13–14
    rendered to clients, collecting money for, 15
    revenue, 104
    solid business model, 75

Short Hair, 132
Short-term goals, 68, 70, 80
Simplified Employee Pension Individual Retirement Account (SEP IRA), 112
Situation analysis, 122
S.M.A.R.T., 68
Smith, Jane, 32
Social media outlets, 130
Sole proprietorship, 31–33
   pros and cons of, 32
Sophisticate's Hairstyle Guide, 132
Spending atmosphere, creating, 85–86
State Board of Cosmetology, 20
Stuff, knowing, 86–87
StyleSync, 130
Sublease agreement, 46–47, 55–60
Success formula, 164–165
Supplies, as tax-deductible category, 112
Supply and demand, 81
S.W.O.T. analysis, 62, 88–91, 122

Taxes, 109–112
   advertising, 110
   estimated, 17
   federal tax ID number, 38–39
   insurance, 110
   interest, 110–111
   and licenses, 112
   meals and entertainment, 112
   mileage, 110
   office expenses, 111
   payments, 106
   rent or lease, 111
   repairs and maintenance, 111
   sales tax ID number, 39
   self-employment, 17
   supplies, 112
   travel, 112
Technical educator, 133
Television commercials, 132
Text marketing, 129
Towel service, 19
Travel, as tax-deductible category, 112
Trend Setters Hair Design Group, 30
Twitter, 130, 149

Vision statement, 67, 73

Wanger, Kristen, 142
Website, 128–129
Weekly financial checklist, 105
Weekly retail tracking form, 171–172
Wholesale cost of goods, 162–163
Work agreement, 46–47, 53–54
Work hours
   scheduling, 98
   setting, 11–12
Work station, 19
www.amtamassage.org, 17
www.ascpskincare.com, 17
www.insuringstyle.com, 17, 108
www.probeauty.org, 17, 108

YouTube, 130